Math Mammoth
Add & Subtract 2-B

By Maria Miller

Contents

Introduction

Math Mammoth Add & Subtract 2-B is a continuation to the book *Math Mammoth Add & Subtract 2-A*. The goal of this book is to study addition and subtraction within 0-100, both mentally and in columns, especially concentrating on regrouping in addition (carrying) and in subtraction (borrowing).

Mental math is important because it builds number sense. This book includes many lessons that practice mental math. For example, the child practices adding and subtracting two-digit numbers when one of the numbers is a whole ten (such as $30 + 14$ or $66 - 20$).

The students also learn to solve sums such as $36 + 8$ or $45 + 9$ with mental math (there is a regrouping), comparing them to the sums $6 + 8$ and $5 + 9$. Since $6 + 8$ fills the first ten and is four more than the next ten, the sum $36 + 8$ fills the *next* whole ten (40), and is four more than that, or 44.

At the same time, students also learn to add two-digit numbers with regrouping. This is explained in detail with the help of visual models (base-ten blocks). You are welcome to use actual physical manipulatives, if you prefer. The main concept to understand is that 10 ones are regrouped to form a new ten, and this new ten is written using a little "1" in the tens column.

As a "stepping stone" into the standard addition algorithm, you can show the child the optional method below. This can be used if the child does not readily understand why the little "1" above the tens column corresponds to a ten. Below, the ones are added first, and the answer is written using both columns. Then, the tens are added and their sum is written under the sum of ones. Lastly, both sums are added.

After addition, the lessons focus on column subtraction, initially without regrouping and then with regrouping.

The lesson *Regrouping* practices breaking down a ten into 10 ones. It is crucial that the child *understands* what happens here. Otherwise, they might end up only memorizing the procedure, and will probably at some point misremember how it was done. If you notice that the child does not understand the concept of regrouping, provide more practice with concrete manipulatives or visual exercises before proceeding.

After learning regrouping, students practice mental subtraction in two separate lessons. One of them expounds on several methods for mental subtracting. The other is about Euclid's game—a fun game that also practices subtraction of two-digit numbers.

I wish you success in teaching math!

Maria Miller, the author

Games and Activities

Make 100

You need: Two standard decks of playing cards from which you remove the face cards and tens, leaving only numbers 1 through 9.

Game Play: In each round, each player is dealt four cards. Each player forms two 2-digit numbers with their four cards, using each card as a digit. For example, if you're dealt 4, 8, 6, and 1, you could make 84 and 16. Or, you could make 41 and 68. The goal is to make these two numbers in such a manner that their sum is as close to 100 as possible. Each player calculates the sum of their numbers <u>mentally</u>. The player with the sum closest to 100 wins that round, and puts all the cards played on that round to their personal pile.

In the case of a tie, the players are dealt four new cards each, and they use those to resolve the tie. After enough rounds have been played to use all of the cards in the deck, the player with the most cards in their personal pile wins.

Variation: Allow players to calculate the sums using pencil and paper. Mental math is much faster, though. (You can always add the tens separately and the ones separately, and add those two sums.)

Simple Dice

You need: five six-sided dice.

The goal of the game to is to get the maximum sum from the five dice. The game practices mental addition of several small numbers.

Game play: At your turn, roll the five dice. You have to leave at least one of the dice (hold it), but you may reroll up to four of them. Again, you have to hold at least one dice, and you can reroll the rest. After these three rolls, your turn is over. Calculate the sum of your dice. This is then written down as your score for this turn.

After a set number of turns (such as five), each player calculates their total score of all the rounds. The player with the highest total wins.

One is IN

This is a variation of the above game, Simple Dice. It adds in one additional rule, and that is why I recommend that you first play the Simple Dice game with your child or students, so they learn the basic idea of the game.

You need: five 6-sided dice

The goal of the game to is to get the maximum sum from the five dice. One of the dice has to show 1, for you to score at all.

Game play: At your turn, roll the five dice. You have to leave at least one of the dice (hold it), but you may reroll up to four of them. Again, you have to hold at least one dice, and you can reroll the rest. After four such rolls, your turn is over. If at least one of your dice shows 1, calculate the sum of your dice. This is then written down as your score for this turn. If none of your dice show 1, you do not score anything.

After a set number of turns (such as five), each player calculates their total score of all the rounds. The player with the highest total wins.

Games and Activities at Math Mammoth Practice Zone

Two-Digit Addition with Mental Math
Simple online practice of adding two-digit numbers using mental math.

- Add a two-digit and a single-digit number:
  https://www.mathmammoth.com/practice/addition-subtraction-two-digit#opts=2p1dwr

- Add two 2-digit numbers, no regrouping:
  https://www.mathmammoth.com/practice/addition-subtraction-two-digit#opts=2p2dnr

- Add two 2-digit numbers, with regrouping:
  https://www.mathmammoth.com/practice/addition-subtraction-two-digit#opts=2p2dwr

Hidden Picture Addition Game
Add two-digit numbers and reveal a hidden picture.
https://www.mathmammoth.com/practice/mystery-picture#min=11&max=99

Mathy's Berry Picking Adventure
The first link practices adding a two-digit and a single-digit number (e.g. 45 + 7). The second link practices mentally adding two 2-digit numbers (e.g. 34 + 26).

- https://www.mathmammoth.com/practice/mathy-berries#mode=addition-both&duration=2m

- https://www.mathmammoth.com/practice/mathy-berries#mode=addition-double&duration=2m

Bingo
For this chapter, choose Addition (Two-Digit) to practice mental addition of two-digit numbers.
https://www.mathmammoth.com/practice/bingo

Fruity Math
Click the fruit with the correct answer and try to get as many points as you can within two minutes.

- Add a two-digit number and nine:
  https://www.mathmammoth.com/practice/fruity-math#op=addition&duration=120&mode=manual&config=12,89x1__9,9x1&max-sum=200

- Add a two-digit and a single-digit number:
  https://www.mathmammoth.com/practice/fruity-math#op=addition&duration=120&mode=manual&config=13,89x1__3,9x1&max-sum=200

- Add two 2-digit numbers:
  https://www.mathmammoth.com/practice/fruity-math#op=addition&duration=120&mode=manual&config=11,89x1__11,99x1&max-sum=125

- Start with single-digit additions, and then advance through levels with increasingly harder sums:
  https://www.mathmammoth.com/practice/fruity-math#op=addition&duration=120&mode=levels&start-level=2

Make number sentences
Drag two flowers to the empty slots to make the given sum, practicing two-digit mental addition.
https://www.mathmammoth.com/practice/number-sentences#questions=5&types=add-11-80

Color-Grid Game — Vertical Addition Practice
Solve 12 problems of adding two-digit numbers in columns.
https://www.mathmammoth.com/practice/vertical-addition#max=99&questions=4*3&addends=2&max-digits=3

Helpful Resources on the Internet

We have compiled a list of external Internet resources that match the topics in this book. This list of links includes web pages that offer:

- **online practice** for concepts;

- online **games**, or occasionally, printable games;

- **animations** and interactive **illustrations** of math concepts;

- **articles** that teach a math concept.

We heartily recommend you take a look at the list. Many of our customers love using these resources to supplement the bookwork. You can use the resources as you see fit for extra practice, to illustrate a concept better, and even just for some fun. Enjoy!

https://l.mathmammoth.com/blue/addsubtract2b

SCAN ME

Adding Within the Same Ten

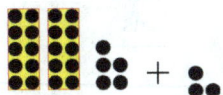

25 + 3 = 28

Add 5 + 3 first.
The 2 tens do not change.

12 + 7 = 19

Add 2 + 7 first.
The ten does not change.

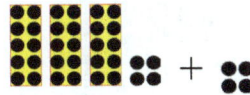

34 + 4 = 38

Add 4 + 4 first.
The 3 tens do not change.

1. Write an addition sentence for each picture.

a.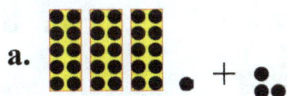

_____ + _____ = _____

b.

_____ + _____ = _____

c.

_____ + _____ = _____

d.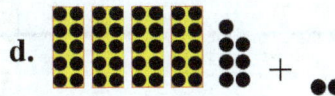

_____ + _____ = _____

2. Add. Compare the problems. The top problem helps you solve the bottom one!

a. 5 + 2 = _____	b. 4 + 5 = _____	c. 3 + 6 = _____
35 + 2 = _____	64 + 5 = _____	93 + 6 = _____

3. Add. For each problem, think about a "helping" problem with numbers less than 10.

a. 52 + 7 = _____	b. 33 + 1 = _____	c. 11 + 5 = _____
_2_ + _7_ = _____	_____ + _____ = _____	_____ + _____ = _____

4. The numbers are written in boxes! Add ones in their own column.
   Copy the tens number down below.

a. $35 + 3$	b. $12 + 6$	c. $57 + 1$	d. $64 + 3$
tens / ones: 3 5, + ↓ 3, = 3 8	tens / ones: 1 2, + ↓ 6	tens / ones: 5 7, + ↓ 1	tens / ones: 6 4, + ↓ 3

5. Now *you* write the numbers in the boxes. Add ones in their own column.

a. $26 + 3$	b. $72 + 4$	c. $65 + 4$	d. $81 + 4$
tens / ones: + ↓	tens / ones: + ↓	tens / ones: + ↓	tens / ones: + ↓

6. Add. Compare the problems.

a.	b.	c.	d.
$6 + 2 =$ _____	$4 + 3 =$ _____	$5 + 4 =$ _____	$11 + 7 =$ _____
$16 + 2 =$ _____	$24 + 3 =$ _____	$45 + 4 =$ _____	$61 + 7 =$ _____
$36 + 2 =$ _____	$34 + 3 =$ _____	$65 + 4 =$ _____	$41 + 7 =$ _____

7. Add many numbers.

a.	b.	c.
$20 + 5 + 2 =$ _____	$93 + 1 + 5 =$ _____	$100 + 5 + 4 =$ _____
$44 + 2 + 2 =$ _____	$83 + 4 + 3 =$ _____	$52 + 4 + 2 =$ _____

Remember how to break a number into its TENS and ONES?	$23 = 20 + 3$ tens ones	$47 = 40 + 7$ tens ones

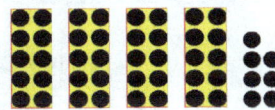

8. Break the numbers into tens and ones or do it the other way around.

a.	b.	c.
$18 = 10 + 8$	$32 = \underline{\quad} + \underline{\quad}$	$\underline{\qquad} = 60 + 6$
$25 = \underline{\quad} + \underline{\quad}$	$95 = \underline{\quad} + \underline{\quad}$	$\underline{\qquad} = 9 + 80$
$55 = \underline{\quad} + \underline{\quad}$	$\underline{\qquad} = 40 + 9$	$\underline{\qquad} = 8 + 70$

9. Compare. Write $<$, $>$, or $=$.

a. $24 + 3 \ \boxed{} \ 24 + 5$	**c.** $17 + 2 \ \boxed{} \ 19 + 2$	**e.** $58 \ \boxed{} \ 8 + 51$
b. $83 + 5 \ \boxed{} \ 85 + 3$	**d.** $36 + 4 \ \boxed{} \ 46 + 4$	**f.** $66 \ \boxed{} \ 5 + 61$

Puzzle Corner ☆ is a number we don't know—a mystery number! Your task is to *compare* without knowing the mystery number! For example, which is more, ☆ + 2 or ☆ + 7?

Write $<$ or $>$ in the boxes. Note: there is one comparison you **cannot** do without knowing the mystery number. Can you find it?

☆ + 5 $\boxed{}$ ☆ + 4   ☆ − 5 $\boxed{}$ ☆ − 4   ☆ − 5 $\boxed{}$ ☆

☆ + 2 $\boxed{}$ ☆ + 7   ☆ − 5 $\boxed{}$ ☆ − 6   ☆ + ☆ $\boxed{}$ ☆ + 20

Subtracting Within the Same Ten

		Think of the *ones digits* only. The tens do not change, because we don't have to subtract from the tens.
14 – 2 = __12__	27 – 3 = __24__	
"I can subtract 4 – 2 = 2; the 10 stays the same."	"I can subtract 7 – 3 = 4; the 20 stays the same."	

1. Subtract and compare. The top problem helps you solve the bottom one!

a. 8 – 2 = __6__ 28 – 2 = __26__	**b.** 7 – 6 = _____ 17 – 6 = _____	**c.** 7 – 7 = _____ 67 – 7 = _____
d. 6 – 6 = _____ 56 – 6 = _____	**e.** 9 – 8 = _____ 49 – 8 = _____	**f.** 5 – 2 = _____ 95 – 2 = _____

2. Subtract. Think of the "helping problem" that only uses numbers less than 10.

a. 54 – 2 = _____ 4 – 2 = _____	**b.** 76 – 2 = _____ ____ – ____ = _____	**c.** 88 – 4 = _____ ____ – ____ = _____

3. Subtract. Cross out dots. The box with "T" is a ten.

a. 35 – 4 = _____ 35 – 3 = _____ 35 – 2 = _____	**b.** 57 – 7 = _____ 57 – 5 = _____ 57 – 3 = _____	**c.** 48 – 2 = _____ 48 – 4 = _____ 48 – 6 = _____	**d.** 34 – 1 = _____ 34 – 2 = _____ 34 – 4 = _____

4. Subtract.

a.	b.	c.	d.
$77 - 6 =$ _____	$47 - 2 =$ _____	$57 - 4 =$ _____	$15 - 3 =$ _____
$22 - 1 =$ _____	$75 - 1 =$ _____	$86 - 2 =$ _____	$98 - 4 =$ _____

5. Find the missing addends.

a. $10 +$ _____ $= 15$	b. $21 +$ _____ $= 22$	c. $65 +$ _____ $= 69$
$32 +$ _____ $= 38$	$94 +$ _____ $= 95$	$33 +$ _____ $= 36$
$72 +$ _____ $= 79$	$44 +$ _____ $= 48$	$91 +$ _____ $= 98$

6. Solve.

a. In the morning, Katherine sold 21 pictures that she had painted and in the afternoon, she sold 7. How many pictures did she sell in total?

b. She had 30 pictures to sell when she started. How many does she have left now?

c. Katherine can paint a picture in one hour. She started painting at 4:30 and painted three pictures. What time did she stop painting?

7. Take away the ones (the dots) so that only the whole tens are left.

a.	b.	c.
$37 -$ _____ $= 30$	$46 -$ _____ $= 40$	$28 -$ _____ $=$ _____
d. $57 -$ _____ $=$ _____	e. $85 -$ _____ $=$ _____	f. $69 -$ _____ $=$ _____

8. Solve. In the last row, make your own problems, and let a friend solve them!

a. $50 + \bigcirc = 57$	**b.** $\bigcirc + 2 = 88$	**c.** $79 - 9 = \bigcirc$
d. $\bigcirc - 5 = 20$	**e.** $90 - \bigcirc = 85$	**f.** $42 = 40 + \bigcirc$
$\bigcirc + \underline{\quad} = \underline{\quad}$		$\underline{\quad} + \bigcirc = \underline{\quad}$

9. Count by fives. Notice the patterns! A 100-chart or an abacus can help you.

   a. 10, 15, _____, _____, _____, _____, _____, _____, _____

   b. 1, 6, _____, _____, _____, _____, _____, _____, _____

   c. 3, 8, _____, _____, _____, _____, _____, _____, _____

10. Continue the patterns.

a.	b.	c.
$88 - 0 = \underline{\quad}$	$95 - 2 = \underline{\quad}$	$48 - 1 = \underline{\quad}$
$88 - 1 = \underline{\quad}$	$85 - 2 = \underline{\quad}$	$46 - 1 = \underline{\quad}$
$88 - 2 = \underline{\quad}$	$75 - 2 = \underline{\quad}$	$44 - 1 = \underline{\quad}$
$88 - \underline{\quad} = \underline{\quad}$	$\underline{\quad} - \underline{\quad} = \underline{\quad}$	$\underline{\quad} - 1 = \underline{\quad}$
$88 - \underline{\quad} = \underline{\quad}$	$\underline{\quad} - \underline{\quad} = \underline{\quad}$	$\underline{\quad} - \underline{\quad} = \underline{\quad}$
$\underline{\quad} - \underline{\quad} = \underline{\quad}$	$\underline{\quad} - \underline{\quad} = \underline{\quad}$	$\underline{\quad} - \underline{\quad} = \underline{\quad}$
$\underline{\quad} - \underline{\quad} = \underline{\quad}$	$\underline{\quad} - \underline{\quad} = \underline{\quad}$	$\underline{\quad} - \underline{\quad} = \underline{\quad}$
$\underline{\quad} - \underline{\quad} = \underline{\quad}$	$\underline{\quad} - \underline{\quad} = \underline{\quad}$	$\underline{\quad} - \underline{\quad} = \underline{\quad}$

Add and Subtract Two-Digit Numbers

1. Subtract. Cover with your fingers, or cross out, what needs to be subtracted.

a. 48 − 20 = _____	**b.** 36 − 30 = _____	**c.** 61 − 50 = _____
d. 55 − 40 = _____	**e.** 44 − 30 = _____	**f.** 72 − 50 = _____

2. Add. You can also use the abacus to help you, instead of the pictures.

a. 35 + 20 = _____	**b.** 21 + 30 = _____	**c.** 47 + 20 = _____
d. 20 + 28 = _____	**e.** 14 + 15 = _____	**f.** 50 + 16 = _____

3. A challenge! Use the 100-bead abacus if you need help.

a.	**b.**	**c.**
35 + 20 = _____	40 + 17 = _____	33 − 20 = _____
76 + 30 = _____	30 + 33 = _____	78 − 50 = _____
22 + 50 = _____	56 − 20 = _____	99 − 40 = _____

We can write the numbers to be added or subtracted UNDER each other in the boxes.

Then we add or subtract the ones in their own column (marked orange).

Then we add or subtract the tens in their own column (marked turquoise).

tens	ones
4	5
+ 2	3
6	8

Which numbers were added?
What is the answer?

tens	ones
7	8
− 5	6
2	2

Which numbers were subtracted?
What is the answer?

4. Add.

a.

tens	ones
4	2
+ 2	4

b.

tens	ones
5	3
+ 0	6

c.

tens	ones
2	5
+ 5	3

d.

tens	ones
3	5
+ 0	4

5. Subtract.

a.

tens	ones
9	5
− 2	0

b.

tens	ones
5	8
− 2	6

c.

tens	ones
2	5
− 0	3

d.

tens	ones
7	9
− 6	4

6. Write the numbers under each other in the boxes and add.

a. 17 + 21

b. 34 + 14

c. 51 + 7

d. 32 + 5

7. Subtract. Either use the picture, or write the numbers under each other in the boxes.

	3	8
−	1	4

a. 38 − 14 = _____

	1	6
−		3

b. 16 − 3 = _____

c. 47 − 25 = _____

d. 38 − 26 = _____

8. The dots show what two numbers you add. Write the numbers in the boxes. Add.

a.

tens	ones
2	4
+ 1	3

b.

tens	ones

c.

tens	ones

d.

tens	ones

e.

tens	ones

f.

tens	ones

9. You should be able to do these problems mentally, without any help!

a. 30 + 20 = _____	b. 40 + 60 = _____	c. 60 − 40 = _____
60 + 20 = _____	30 + 30 = _____	70 − 50 = _____
30 + 50 = _____	50 − 20 = _____	90 − 40 = _____

10. Write the numbers in the boxes. Subtract the tens and the ones in their columns.

a. 57 – 21   **b.** 74 – 14   **c.** 59 – 7   **d.** 99 – 58

11. Solve the word problems. You can use an abacus to help, or write the numbers under each other in boxes.

a. James went fishing and caught 28 fish. His wife cooked 11 fish for supper. How many fish were not used for supper, and were put into the freezer?

b. If you buy a shirt for $22 and jeans for $34, how much do you have to pay for both items?

c. Mom is 38 years old and John is 11. How many years older is Mom than John?

d. Jessica had 34 colored pencils and Matt had 22. Jessica gave Matt 6 pencils. Now how many does Matt have?

18

Subtract from Whole Tens

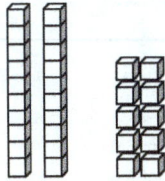

1. Subtract from whole tens. The last ten is shown with blocks so you can cross some out!

a.	b.	c.	d.
40 − 4 = _____	30 − 5 = _____	50 − 2 = _____	60 − 7 = _____
40 − 6 = _____	30 − 4 = _____	50 − 8 = _____	60 − 9 = _____
40 − 7 = _____	30 − 9 = _____	50 − 3 = _____	60 − 1 = _____
40 − 8 = _____	30 − 6 = _____	50 − 6 = _____	60 − 4 = _____

2. Subtract the same number many times.

a. 70 − 10 − 10 − 10 = _____	b. 90 − 20 − 20 − 20 = _____

3. Add and subtract whole tens.

10   ↗ +50   ↘ +40   ↗ −20   ↘ −60   ↗ +20   ↘ −30   ↗ +20   ↘ +60   ↗ −30

4. Subtract from whole tens.

a.	b.	c.	d.
$70 - 6 =$ _____	$50 - 8 =$ _____	$40 - 1 =$ _____	$100 - 5 =$ _____
$70 - 5 =$ _____	$50 - 7 =$ _____	$40 - 2 =$ _____	$100 - 7 =$ _____
$70 - 2 =$ _____	$50 - 6 =$ _____	$40 - 3 =$ _____	$100 - 9 =$ _____

5. Add and subtract the same number.

a. $10 - 2 =$ _____	b. $60 - 5 =$ _____	c. $25 - 4 =$ _____
$10 + 2 =$ _____	$60 + 5 =$ _____	$25 + 4 =$ _____

6. Write an addition or subtraction sentence for each problem, and solve it.

a. There are 20 pupils in the class, and each needed
a pencil. The teacher only found 16 pencils.
How many pencils do they still need?

b. There are 17 bushes growing in the front yard
and seven in the back yard. How many more are
in the front yard than in the back yard?

c. Carmen has 13 pretty stones, Jane has 18, and Julie has 20.

How many more stones does Julie have than Carmen?

How many more stones does Jane have than Carmen?

How many more stones does Carmen need if she
wants to have as many as Julie?

Doubling

Doubling a number means adding it to itself. It is finding two times the number.

Examples:

Double 7 is 7 + 7 is <u>14</u>.   Double 20 is 20 + 20 is <u>40</u>.

1. Find the double of these numbers.

a. Double 4	**b.** Double 6	**c.** Double 8
_____ + _____ = _____	_____ + _____ = _____	_____ + _____ = _____
d. Double 10	**e.** Double 30	**f.** Double 50
_____ + _____ = _____	_____ + _____ = _____	_____ + _____ = _____

2. Find the double of these numbers by adding in the boxes.

　　a. 22 + 22　　　　**b.** Double 34　　　**c.** 13 + 13　　　**d.** Double 41

3. Make a doubles chart. Notice it has a pattern!

Double 1 = _____	6 + 6 = _____	11 + 11 = _____
Double 2 = _____	7 + 7 = _____	12 + 12 = _____
Double 3 = _____	8 + 8 = _____	13 + 13 = _____
Double 4 = _____	9 + 9 = _____	14 + 14 = _____
Double 5 = _____	10 + 10 = _____	15 + 15 = _____

When you double a number, you always get an EVEN number as a result.

Look at the ANSWER numbers in the doubles chart you just made. (You can color the answers yellow if you would like.)

All of those numbers are EVEN numbers.

If a number is even, you can share that many things evenly.

Example. Double 13 is $13 + 13 = 26$. This means that two children can share 26 toy cars EVENLY, and that each child gets 13 cars.

Example. Two children need to clean 18 chairs. They divide the job equally (evenly). How many chairs does each child clean?

Use the doubles chart. Since $9 + 9 = 18$, each child will clean 9 chairs.

Solve the problems. You can use the doubles chart on the previous page to help you.

4. Mother tells two children to make 16 sandwiches.
   The children share the job equally.
   How many sandwiches will each child make?

5. You have 12 grapes and you share them evenly with your
   sister. How many do you get?

6. In a board game, you throw two dice and move that many spaces.
   Mary got double four. Andrea got double six.
   How many spaces did Mary move?

   How many spaces did Andrea move?

7. Cindy has 5 apples and Sandy has 3. They put them together
   and share them evenly. How many does each girl get?

8. Circle the even numbers: 13   20   19   8   15   16

Each number here is an even number, so it is a DOUBLE of some number. What number is it double of?

6	8	10	12	14	16	18	20	22	24

The first number on the list is 6. Six is double 3. We can write 6 = 3 + 3.
The last number on the list is 24. It is double 12. We can write 24 = 12 + 12.

9. Write each number as a double of some other number.

a. 8 = ____ + ____	b. 10 = ____ + ____	c. 4 = ____ + ____
d. 12 = ____ + ____	e. 14 = ____ + ____	f. 16 = ____ + ____

10. Write above each shaded number what number it is double of. Notice the pattern!

		5										
6	8	10	12	14	16	18	20	22	24	26	28	30

11. Mother and her friend need to make 20 dolls to sell.
They share the job evenly.
How many dolls will each woman make?

12. Two teachers divide 28 worksheets evenly.
How many worksheets will each one get?

13. Mom found 7 cucumber slices in one container and 3 in another.
You and your brother decide to share them equally.
How many slices will you get?

14. (Challenge) A batch of brownies makes 16 brownies.
Mom makes a double batch.
How many brownies will she make?

One-Half

<table>
<tr>
<td>

If you divide something into *two equal* parts, you have divided it into two halves. Each part is **half** of the whole.

Write one-half this way: $\frac{1}{2}$, or this way: 1/2.

</td>
<td>

You can also find **half** of so many **things**, if you have an <u>even</u> number of things.

5 + 5 = 10. So, half of ten apples is five apples.

</td>
<td>

Twelve balls are divided into two equal parts. We can do that, because 12 is an even number.

6 + 6 = 12

$\frac{1}{2}$ of 12 is 6.

</td>
</tr>
</table>

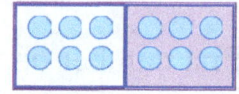

1. **a.** Color one half of each shape.   **b.** Color <u>two</u> halves of each shape.

2. Draw a line through these shapes and divide them into two halves. Color one half.

   a.   **b.**   **c.**   **d.**   **e.**

3. Divide the things into two EQUAL groups. Write an addition. Find half of the total.

<table>
<tr>
<td>

a. 10

____ + ____ = ____

$\frac{1}{2}$ of 10 is ____.

</td>
<td>

b. 40

____ + ____ = ____

$\frac{1}{2}$ of 40 is ____.

</td>
<td>

c. 24

____ + ____ = ____

$\frac{1}{2}$ of 24 is ____.

</td>
</tr>
</table>

Doubling and halving are the opposite things. $7 + 7 = 14$, so $\frac{1}{2}$ of 14 is 7.

4. Fill in the doubles chart. Then use it to find one-half of the given numbers.

6 + 6 = _____	11 + 11 = _____
7 + 7 = _____	12 + 12 = _____
8 + 8 = _____	13 + 13 = _____
9 + 9 = _____	14 + 14 = _____
10 + 10 = _____	15 + 15 = _____

$\frac{1}{2}$ of 16 is _____.

$\frac{1}{2}$ of 28 is _____.

$\frac{1}{2}$ of 26 is _____.

$\frac{1}{2}$ of 30 is _____.

$\frac{1}{2}$ of 22 is _____.

5. Divide the dots into two EQUAL groups. Find half of the total.

a. $\frac{1}{2}$ of 30 is _____.

b. $\frac{1}{2}$ of _____ is _____.

c. $\frac{1}{2}$ of _____ is _____.

6. Solve the problems. Then fill in another chart of doubles. *It has a pattern!* Find it!

a. Jack and Joe split $60 evenly.
   How many dollars did each one get?

b. Half of 100 students were sick.
   How many were not sick?

c. Aunt Katie gave Missie half of $40.
   Missie spent $10 on a toy.
   How many dollars does Missie have now?

d. The recipe called for 10 apples. That was exactly half of Mom's apples. How many apples did Mom have in the beginning?

10 + 10 = _____
15 + 15 = _____
20 + 20 = _____
25 + 25 = _____
30 + 30 = _____
35 + 35 = _____
40 + 40 = _____

Pictographs

1. This is a **pictograph.** It shows how many miles each boy has ridden on his bike. **Each bicycle** picture means **10 miles**. A half-bicycle would be half that.

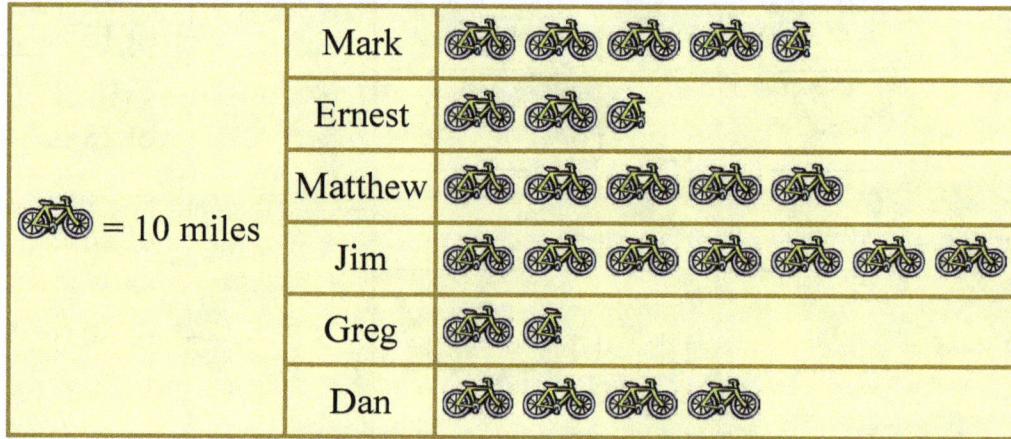

	Mark	🚲 🚲 🚲 🚲 🚲
	Ernest	🚲 🚲 🚲
🚲 = 10 miles	Matthew	🚲 🚲 🚲 🚲 🚲
	Jim	🚲 🚲 🚲 🚲 🚲 🚲 🚲
	Greg	🚲 🚲
	Dan	🚲 🚲 🚲 🚲

  a. Who rode the most miles? _____

  How many miles? _____ miles

  b. The boys that rode the least miles were Greg and _____.

  How many miles did they ride? _____ and _____ miles

  c. How many more miles did Matthew ride than Dan? _____ miles

  d. How many more miles did Dan ride than Greg? _____ miles

2. Make a bar graph.

Favorite color	How many people like it
Blue	30
Green	17
Red	13
Purple	12
Black	10

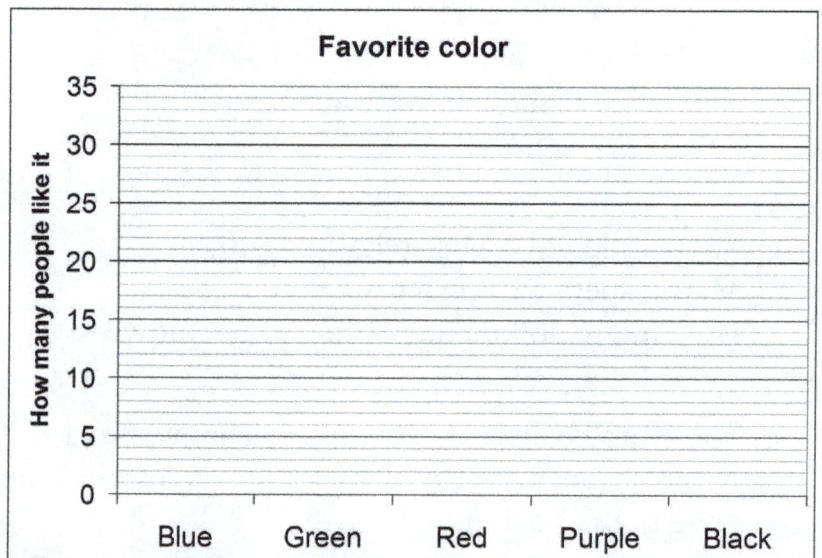

Favorite color

How many people like it

35
30
25
20
15
10
5
0

Blue   Green   Red   Purple   Black

3. The children picked fruit from grandpa Jerry's fruit trees. **One** fruit **picture** means **6 fruits**. Half a fruit means 3 fruits.

	How many?	
oranges		
mangos		
bananas		

a. Fill in how many fruits they picked.

b. How many oranges and bananas did they pick?
You can use the grid on the right to add.

c. How many more bananas did they pick than mangos?

tens	ones
+	

4. Boys played a game of marbles and made a pictograph. Each marble means 5 points.

Mark	
Aaron	
Henry	
Jack	

Make three questions that you could ask another second grader about this pictograph. Then, ask the questions of your classmate or a friend. Check their answers.

Adding with Whole Tens

1. The numbers are shown with ten-sticks and one-dots. Write the sums.

a. $54 + 10 = $ _____

b. _____ $+ 20 = $ _____

c. _____ $+$ _____ $=$ _____

d. _____ $+$ _____ $=$ _____

e. _____ $+$ _____ $=$ _____

f. _____ $+$ _____ $=$ _____

Adding whole tens and another 2-digit number	$50 + 26$	$39 + 40$
Break down the other number into tens and ones. Add the tens. Then, add the ones.	$50 + 20 + 6$ $70 + 6 = 76$	$30 + 9 + 40$ $70 + 9 = 79$

2. Add. Break the second number into tens and ones first. Then add the tens.

a. $10 + \underline{34} = $ _____ $(10 + \underline{30} + \underline{4})$	b. $10 + 28 = $ _____ $(10 + $ ___ $ + $ ___ $)$	c. $20 + 24 = $ _____ $(20 + $ ___ $ + $ ___ $)$
d. $30 + 21 = $ _____	e. $50 + 17 = $ _____	f. $40 + 33 = $ _____
g. $60 + 23 = $ _____	h. $30 + 37 = $ _____	i. $70 + 25 = $ _____

3. Add. Break the first number into tens and ones first. Then add the tens.

a. $45 + 20 =$ _____ $(\underline{40} + \underline{5} + 20)$	**b.** $27 + 20 =$ _____ $(\_\_\_ + \_\_\_ + 20)$	**c.** $45 + 40 =$ _____ $(\_\_\_ + \_\_\_ + 40)$
d. $46 + 30 =$ _____	**e.** $16 + 50 =$ _____	**f.** $38 + 60 =$ _____
g. $20 + 77 =$ _____	**h.** $58 + 40 =$ _____	**i.** $40 + 39 =$ _____

4. Explain in your own words how you can mentally add $21 + 60$.

5. Fill in the chart of doubles again, and notice its PATTERN.

$5 + 5 =$ _____	$30 + 30 =$ _____
$10 + 10 =$ _____	$35 + 35 =$ _____
$15 + 15 =$ _____	$40 + 40 =$ _____
$20 + 20 =$ _____	$45 + 45 =$ _____
$25 + 25 =$ _____	$50 + 50 =$ _____

6. Isabella got 30 books out of the library, and read half of them in two days. How many books does she have left to read?

7. Gwen and Mom went shopping. They bought shoes for $40, a top for $10, and a skirt for $20. Mom paid half of the cost and Gwen paid the rest. How much did Gwen pay?

8. Jacob had $61. Then he bought a toy for $30. How much money does he have left?

9. Fill in the missing numbers and find how many tens were added.

a. $12 + \underline{\hspace{2cm}} = 22$	**b.** $45 + \underline{\hspace{2cm}} = 65$	**c.** $23 + \underline{\hspace{2cm}} = 63$
$12 + \underline{\hspace{2cm}} = 52$	$45 + \underline{\hspace{2cm}} = 55$	$23 + \underline{\hspace{2cm}} = 53$
$12 + \underline{\hspace{2cm}} = 42$	$45 + \underline{\hspace{2cm}} = 75$	$23 + \underline{\hspace{2cm}} = 93$

10. Add 10, 20, 30, or 40. In the box below the number, write "E" if the number is even, and "O", if the number is odd. What do you notice?

+ 10

12	22
E	E

+ 20

19	
O	

+ 30

32	

+ 40

37	

+ 40

23	

+ 30

58	

+ 20

7	

+ 10

85	

Puzzle Corner

How many different solutions can you find for this puzzle? Find at least two. All numbers are whole tens.

	+		+		=	70
+	■	+	■	+		
	+		+		=	100
+	■	+	■	+		
	+		+		=	70
=		=		=		
80		100		60		

	+		+		=	70
+	■	+	■	+		
	+		+		=	100
+	■	+	■	+		
	+		+		=	70
=		=		=		
80		100		60		

Subtracting Whole Tens

In the problem 47 − 20, think of the *tens*. The first number (47) has four tens. We take away two tens. So, there are TWO tens left.

The first number also has 7. That does not change.

Cross out two tens.

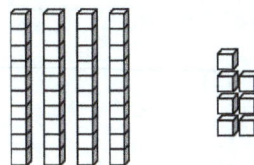

$$47 - 20 = \underline{\hspace{2cm}}$$

1. Cross out as many ten-pillars as the problem indicates. What is left?

a. $70 - 50 = \underline{\hspace{1.5cm}}$

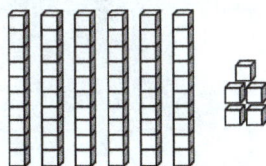

b. $65 - 30 = \underline{\hspace{1.5cm}}$

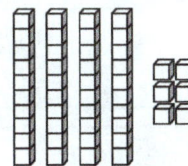

c. $46 - 20 = \underline{\hspace{1.5cm}}$

Notice: The amount of ONES does not change in these subtractions. You can just think of the TENS.

2. Count by tens backwards.

a. 76, 66, \underline{\hspace{1.5cm}} , \underline{\hspace{1.5cm}} , \underline{\hspace{1.5cm}} , \underline{\hspace{1.5cm}} , \underline{\hspace{1.5cm}}

b. \underline{\hspace{1.5cm}} , \underline{\hspace{1.5cm}} , 52, 42, \underline{\hspace{1.5cm}} , \underline{\hspace{1.5cm}} , \underline{\hspace{1.5cm}}

3. Subtract.

a.	b.	c.
$23 - 10 = \underline{\hspace{1.5cm}}$	$48 - 20 = \underline{\hspace{1.5cm}}$	$56 - 10 = \underline{\hspace{1.5cm}}$
$23 - 20 = \underline{\hspace{1.5cm}}$	$48 - 30 = \underline{\hspace{1.5cm}}$	$56 - 30 = \underline{\hspace{1.5cm}}$
d.	**e.**	**f.**
$75 - 10 = \underline{\hspace{1.5cm}}$	$31 - 10 = \underline{\hspace{1.5cm}}$	$81 - 40 = \underline{\hspace{1.5cm}}$
$75 - 20 = \underline{\hspace{1.5cm}}$	$31 - 20 = \underline{\hspace{1.5cm}}$	$81 - 50 = \underline{\hspace{1.5cm}}$

4. Find the pattern and continue it.

a. $88 - 10 = $ _____	**b.** $100 - 60 = $ _____	**c.** $34 - 10 = $ _____
$88 - 20 = $ _____	$90 - 50 = $ _____	$44 - 20 = $ _____
$88 - 30 = $ _____	$80 - 40 = $ _____	$54 - 30 = $ _____
$88 - $ _____ $= $ _____	_____ $- $ _____ $= $ _____	_____ $- $ _____ $= $ _____
$88 - $ _____ $= $ _____	_____ $- $ _____ $= $ _____	_____ $- $ _____ $= $ _____
$88 - $ _____ $= $ _____	_____ $- $ _____ $= $ _____	_____ $- $ _____ $= $ _____
$88 - $ _____ $= $ _____	_____ $- $ _____ $= $ _____	_____ $- $ _____ $= $ _____

5. Solve.

   a. Three suitcases weigh 30 kg, 18 kg, and 20 kg.
   How much is their total weight?

   b. Chairs cost $30 apiece. Can
   Dale buy three of them with $80?

   c. Henry received $50 for his birthday.
   If he buys three books that cost $10 each,
   how much will he have left?

Puzzle Corner

Find numbers for the puzzles.

Completing the Next Ten

Review:			
Review: What numbers make 10? You need to remember these well!	$1 +$ _____ $= 10$	$8 +$ _____ $= 10$	$3 +$ _____ $= 10$
	$7 +$ _____ $= 10$	$5 +$ _____ $= 10$	$6 +$ _____ $= 10$
	$4 +$ _____ $= 10$	$9 +$ _____ $= 10$	$2 +$ _____ $= 10$

Completing the ten

$11 + 9 = 20$

The 1 and the 9 little cubes make a *new* ten. We get a total of 20.

$31 + 9 = 40$

The 1 and the 9 little cubes make a *new* ten. We get a total of 40.

1. Draw more little blocks so you have ten of them. Circle the ten little blocks.
   Write an addition that completes the next ten. You can also do these with the abacus.

a. $33 +$ _____ $=$ _40_

b. $43 +$ _____ $=$ _____

c. $27 +$ _____ $=$ _____

d.   _____ $+$ _____ $=$ _____

e.   _____ $+$ _____ $=$ _____

f.   _____ $+$ _____ $=$ _____

33

2. Write the previous and next whole ten.

a. __10__ , 13 , __20__	b. _____ , 57 , _____	c. _____ , 46 , _____
d. _____ , 81 , _____	e. _____ , 78, _____	f. _____ , 94 , _____

3. Draw more little blocks so you have ten of them. Circle the ten little blocks. Write an addition that completes the next ten. You can also do these with the abacus.

a. _____ + _____ = _____

b. _____ + _____ = _____

c. _____ + _____ = _____

4. Complete the next ten. The top problem is a "helping problem" for the bottom one.

a. $3 + \underline{\quad} = 10$	b. $4 + \underline{\quad} = 10$	c. $7 + \underline{\quad} = 10$
$23 + \underline{\quad} = 30$	$44 + \underline{\quad} = \underline{\quad}$	$17 + \underline{\quad} = \underline{\quad}$

5. Complete the next ten. Think of the helping problem where you complete 10.

a. $13 + \bigcirc = 20$	b. $21 + \bigcirc = 30$	c. $74 + \bigcirc = 80$
d. $88 + \bigcirc = 90$	e. $44 + \bigcirc = 50$	f. $96 + \bigcirc = 100$
g. $\bigcirc + 37 = 40$	h. $\bigcirc + 65 = 70$	i. $\bigcirc + 52 = 60$
j. $\bigcirc + 68 = 70$	k. $\bigcirc + 91 = 100$	l. $\bigcirc + 59 = 60$

6. Complete the next ten. Then write a matching subtraction using the same numbers.
   Notice: The number in the oval will be the same for the top and bottom problem!

a.	**b.**	**c.**
$36 + \bigcirc = 40$	$57 + \bigcirc = \underline{\quad}$	$83 + \bigcirc = \underline{\quad}$
$40 - \bigcirc = 36$	$60 - \bigcirc = \underline{\quad}$	$\underline{\quad} - \bigcirc = \underline{\quad}$

d. $66 + \bigcirc = \underline{\quad}$

$\underline{\quad} - \bigcirc = \underline{\quad}$

e. $95 + \bigcirc = \underline{\quad}$

$\underline{\quad} - \bigcirc = \underline{\quad}$

7. A ticket to an amusement park costs $40. How much *more money* do these children
   need for the ticket?

   a. Jeanine's mom gave her $30 for the ticket, and she has $7.

   b. Derek already has $20. His parents will pay $10.

   c. Muhammad has $12, and his mom has promised him $20.

Puzzle Corner Find two different solutions to the puzzle.

100	−		−		= 40
−		+		+	
	+		+		= 90
=		=		=	
70		40		80	

100	−		−		= 40
−		+		+	
	+		+		= 90
=		=		=	
70		40		80	

35

Going Over Ten

Remember? 10 plus 3, 4, 5, 6, 7, 8, or 9 makes one of the **TEEN** numbers!	10 plus <u>three</u> is <u>thirteen</u>. 10 plus <u>six</u> is <u>sixteen</u>. 10 plus <u>nine</u> is <u>nineteen</u>. 10 plus <u>five</u> is <u>fifteen</u>.

1. Add.

a. $10 + 4 =$ _____   **b.** $10 + 7 =$ _____   **c.** $10 + 8 =$ _____   **d.** $10 + 3 =$ _____

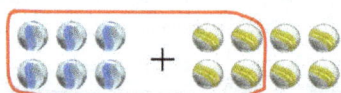 $\quad$ 6 $\quad$ + $\quad$ 8 We circle TEN marbles to make a ten. We can now see that there are 10 and 4 marbles. $10 + 4 = 14$. So, $6 + 8 = 14$.	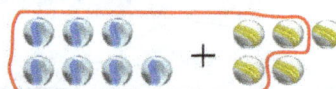 $\quad$ 7 $\quad$ + $\quad$ 5 We circle TEN marbles to make a ten. We can now see that there are 10 and 2 marbles. $10 + 2 = 12$. So, $7 + 5 = 12$.

2. First, circle ten marbles to make a ten. How many marbles are there in all?

a.  $\quad$ 7 $\quad$ + $\quad$ 8 = _____	**b.**  $\quad$ 8 $\quad$ + $\quad$ 8 = _____	**c.**  $\quad$ 6 $\quad$ + $\quad$ 5 = _____
d.  $\quad$ 9 $\quad$ + $\quad$ 4 = _____	**e.**  $\quad$ 8 $\quad$ + $\quad$ 5 = _____	**f.**  $\quad$ 8 $\quad$ + $\quad$ 9 = _____
g.  $\quad$ 7 $\quad$ + $\quad$ 7 $\quad$ = _____	**h.**  $\quad$ 9 $\quad$ + $\quad$ 9 $\quad$ = _____	

Sums that go over to the next ten

Let's add 59 + 5. *First* we complete 60.

59 + 5
$\quad\quad$| $\quad$\
59 + 1 + 4

$\quad$**60** + 4 = 64

The 5 is broken into two parts: 1 and 4.
That is because 59 and 1 makes sixty.
Then, we have 60 and 4. We get 64.

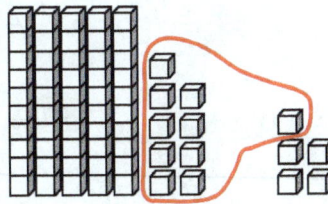

9 and 1 make a new ten.
We get 6 tens.

59 + 5 = 64

---

3. Circle ten little cubes to make a ten. Count the tens and ones. Write the answer.

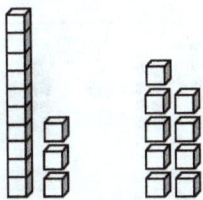

a. 13 + 9 = _____

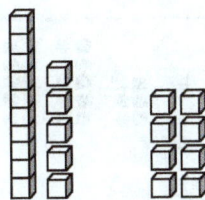

b. 15 + 8 = _____

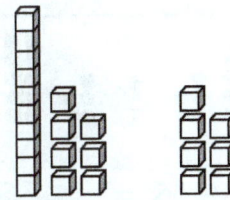

c. 17 + 7 = _____

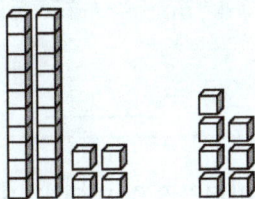

d. 24 + 7 = _____

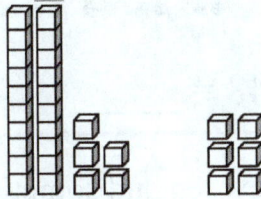

e. 25 + 6 = _____

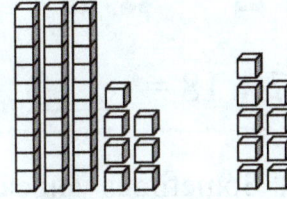

f. 37 + 9 = _____

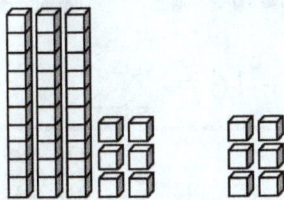

g. 36 + 6 = _____

h. 48 + 4 = _____

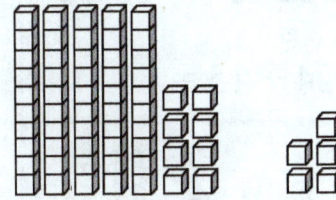

i. 58 + 5 = _____

4. Complete. Break the second number into two parts so that you first complete the next
   ten. You can also use your abacus to solve these.

a. 28 + 8
   / \
 28 + _2_ + _6_
 	
 30 + ___ = _____
	b. 47 + 5
	   / \
	 47 + _3_ + ___
	
	 50 + ___ = _____
		c. 79 + 9
		   / \
		 79 + ___ + ___
		
		 80 + ___ = _____

d. 39 + 3
   / \
 39 + ___ + ___
 	
 40 + ___ = _____
	e. 27 + 5
	   / \
	 27 + ___ + ___
	
	 _____ + ___ = _____
		f. 38 + 7
		   / \
		 38 + ___ + ___
		
		 _____ + ___ = _____

5. Add. First, make a new ten with some of the little dots. You can also use the abacus.

a. 25 + 15 = _____	**b.** 17 + 25 = _____	**c.** 38 + 26 = _____
d. 17 + 18 = _____	**e.** 14 + 48 = _____	**f.** 37 + 24 = _____

6. Add. Sometimes you can make a new ten and sometimes not. An abacus can help also.

a. 24 + 15 = _____	**b.** 18 + 22 = _____	**c.** 36 + 16 = _____
d. 15 + 23 = _____	**e.** 43 + 16 = _____	**f.** 25 + 37 = _____

7. The family counted how many birds they saw on their trip. They used tally marks.

		Count	
Dad	⃦⃦⃦⃦⃦ ⃦⃦⃦⃦⃦ ⃦⃦⃦⃦⃦ IIII		
Mom	⃦⃦⃦⃦⃦ ⃦⃦⃦⃦⃦ ⃦⃦⃦⃦⃦ ⃦⃦⃦⃦⃦ ⃦⃦⃦⃦⃦ III		
Mary	⃦⃦⃦⃦⃦ ⃦⃦⃦⃦⃦ II		
Mark	⃦⃦⃦⃦⃦ ⃦⃦⃦⃦⃦ ⃦⃦⃦⃦⃦ ⃦⃦⃦⃦⃦ ⃦⃦⃦⃦⃦		
Angie	⃦⃦⃦⃦⃦ ⃦⃦⃦⃦⃦ ⃦⃦⃦⃦⃦ ⃦⃦⃦⃦⃦ ⃦⃦⃦⃦⃦ ⃦⃦⃦⃦⃦ ⃦⃦⃦⃦⃦ I		

a. Fill in the *Count* column in the chart.

b. Make a bar graph (below).

c. How many more birds did Dad see than Mary?

d. How many more birds did Angie see than Mark?
   Use subtraction. Write the numbers under each other.

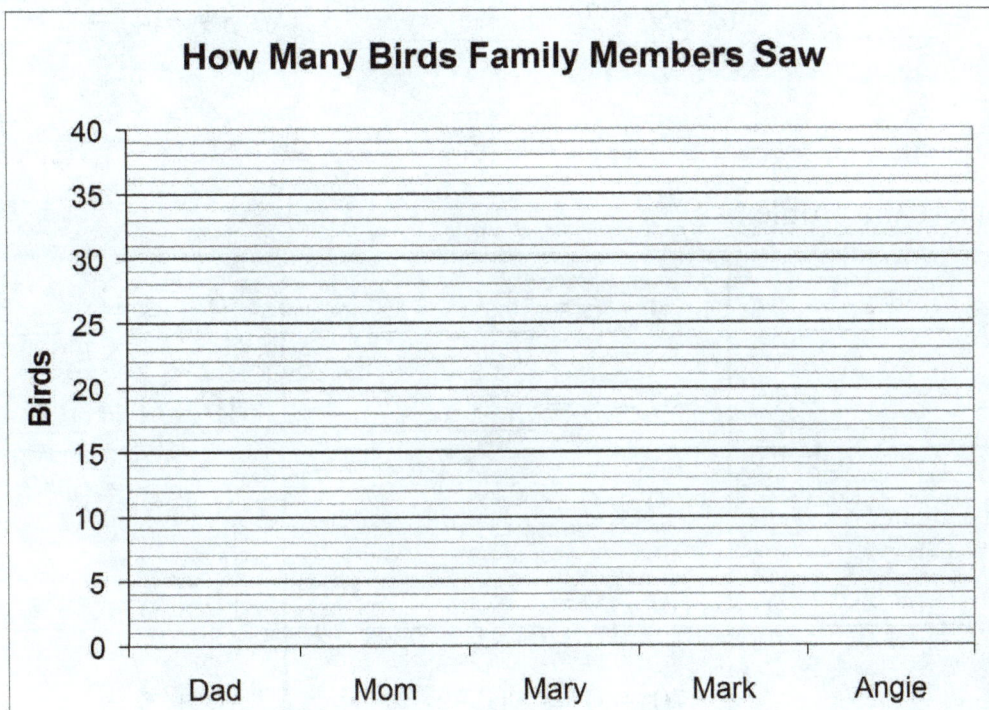

How Many Birds Family Members Saw

Birds

40
35
30
25
20
15
10
5
0

Dad   Mom   Mary   Mark   Angie

Add with Two-Digit Numbers Ending in 9

Imagine that 29 wants to be 30...
so it "grabs" one from 5.
Then, 29 becomes 30, and 5 becomes 4.

The addition problem is changed to 30 + 4 = 34.

$$29 + 5 = \underline{\hspace{2cm}}$$

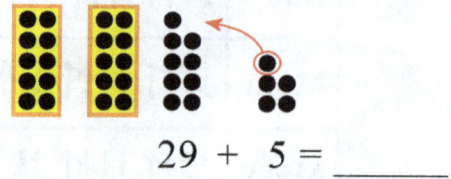

1. Circle the nine dots and one more dot to form a complete ten. Add.

a. $19 + 5 = \underline{\hspace{1.5cm}}$

b. $29 + 7 = \underline{\hspace{1.5cm}}$

c. $49 + 5 = \underline{\hspace{1.5cm}}$

d. $29 + 8 = \underline{\hspace{1.5cm}}$

e. $39 + 6 = \underline{\hspace{1.5cm}}$

f. $49 + 9 = \underline{\hspace{1.5cm}}$

2. Add. Write a helping problem using the "ones" from the first problem.

a. $19 + 7 = \underline{\hspace{1.5cm}}$

$\underline{\ 9\ } + \underline{\ 7\ } = \underline{\hspace{1cm}}$

b. $49 + 3 = \underline{\hspace{1.5cm}}$

$\underline{\hspace{1cm}} + \underline{\hspace{1cm}} = \underline{\hspace{1cm}}$

c. $39 + 4 = \underline{\hspace{1.5cm}}$

$\underline{\hspace{1cm}} + \underline{\hspace{1cm}} = \underline{\hspace{1cm}}$

3. Add. Compare the problems.

a. $9 + 3 = \underline{\hspace{1.5cm}}$

$19 + 3 = \underline{\hspace{1.5cm}}$

b. $9 + 6 = \underline{\hspace{1.5cm}}$

$39 + 6 = \underline{\hspace{1.5cm}}$

c. $9 + 4 = \underline{\hspace{1.5cm}}$

$49 + 4 = \underline{\hspace{1.5cm}}$

d. $9 + 7 = \underline{\hspace{1.5cm}}$

$39 + 7 = \underline{\hspace{1.5cm}}$

$29 + 7 = \underline{\hspace{1.5cm}}$

e. $9 + 9 = \underline{\hspace{1.5cm}}$

$69 + 9 = \underline{\hspace{1.5cm}}$

$79 + 9 = \underline{\hspace{1.5cm}}$

f. $9 + 5 = \underline{\hspace{1.5cm}}$

$19 + 5 = \underline{\hspace{1.5cm}}$

$59 + 5 = \underline{\hspace{1.5cm}}$

4. These problems review the basic facts with 9 and 8. By this time you should already
   remember these addition facts. Try to remember what number will fit without
   counting.

a.	b.	c.	d.
$14 - 9 =$ _____	$4 + 9 =$ _____	$15 -$ _____ $= 8$	$7 + 8 =$ _____
$15 - 9 =$ _____	$8 + 9 =$ _____	$17 -$ _____ $= 8$	$5 + 8 =$ _____
$13 - 9 =$ _____	$5 + 9 =$ _____	$12 -$ _____ $= 8$	$6 + 8 =$ _____
$18 - 9 =$ _____	$6 + 9 =$ _____	$14 -$ _____ $= 8$	$3 + 8 =$ _____
$17 - 9 =$ _____	$9 + 9 =$ _____	$13 -$ _____ $= 8$	$9 + 8 =$ _____
$16 - 9 =$ _____	$7 + 9 =$ _____	$16 -$ _____ $= 8$	$4 + 8 =$ _____

5. Find the difference of numbers. The number line can help.

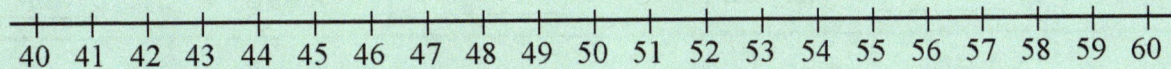

40 41 42 43 44 45 46 47 48 49 50 51 52 53 54 55 56 57 58 59 60

a. Difference between	b. Difference between	c. Difference between
41 and 53 _____	60 and 46 _____	59 and 48 _____

6. Find the patterns and continue them!

a.
$+\boxed{}$ $+\boxed{}$ $+\boxed{}$ $+\boxed{}$ $+\boxed{}$ $+\boxed{}$ $+\boxed{}$ $+\boxed{}$

0   1   3   6   10   _____ _____ _____ _____

b.
$+\boxed{}$ $+\boxed{}$ $+\boxed{}$ $+\boxed{}$ $+\boxed{}$ $+\boxed{}$ $+\boxed{}$ $+\boxed{}$

_____ _____ _____ _____ _____ 44   48   52   56

Add a Two-Digit Number and a Single-Digit Number Mentally

Imagine that 38 wants to be 40, so it "grabs" two from 7. Then, 38 becomes 40, and 7 becomes 5.

The addition problem is changed to 40 + 5 = 45.

38 + 7 = _____

1. Circle the eight dots and two more dots to form a complete ten. Add.

a. 18 + 6 = _____

b. 28 + 7 = _____

c. 48 + 8 = _____

d. 38 + 4 = _____

e. 38 + 6 = _____

f. 48 + 5 = _____

2. Add. Think of the trick explained above.

a. 18 + 7 = _____

b. 38 + 6 = _____

c. 58 + 5 = _____

3. Add. Compare the problems. How are the problems in each box similar?

a. 8 + 3 = _____

18 + 3 = _____

b. 8 + 6 = _____

38 + 6 = _____

c. 8 + 4 = _____

78 + 4 = _____

d. 8 + 2 = _____

38 + 2 = _____

28 + 2 = _____

e. 8 + 9 = _____

68 + 9 = _____

78 + 9 = _____

f. 8 + 5 = _____

18 + 5 = _____

58 + 5 = _____

When you add a two-digit number and a single-digit number, such as 45 + 6 or 77 + 4, think of the "helping" problem: the addition with just the ones digits.

Example. 45 + 6	Example. 67 + 8
Think of the helping problem 5 + 6 = 11. (Drop the 40 from 45, and you have 5 + 6.) 5 + 6 is ONE more than the next ten (11), and 45 + 6 is also ONE more than the next ten (51).	Think of the helping problem 7 + 8 = 15. (Drop the 60 from 67, and you have 7 + 8.) 7 + 8 is FIVE more than the next ten (15), and 67 + 8 is also FIVE more than the next ten (75).

4. Add. Compare the problems! The top problem is a helping problem for the bottom one.

a. 7 + 6 = _____ 27 + 6 = _____ (three more than the next ten)	b. 6 + 8 = _____ 76 + 8 = _____ (four more than the next ten)	c. 7 + 7 = _____ 87 + 7 = _____ (four more than the next ten)
d. 5 + 8 = _____ 35 + 8 = _____	e. 6 + 9 = _____ 26 + 9 = _____	f. 8 + 7 = _____ 48 + 7 = _____

5. Fill in: To add 73 + 8, I can use the helping problem ___ + ___ = ____. Then since

the answer to that is ___ more than 10, the answer to 73 + 8 is ____ more than ____.

6. Add.

a. 34 + 8 = _____	b. 47 + 7 = _____	c. 59 + 5 = _____

7. Solve the word problems.

a. Jenny needed 24 eggs to make omelets for her family. She already had 10 eggs. How many more does she need?

b. Her large family eats lots of potatoes. Dad bought a 25-kilogram bag of potatoes. Now, only 5 kg are left. How many kilograms of potatoes have they eaten?

Regrouping with Tens

When adding 3 + 9, we can circle ten little ones to form a ten. We write "1" in the tens column.

There are two little ones left over, so we write "2" in the ones column.

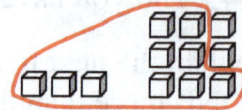

	tens	ones
		3
+		9
	1	2

With 35 + 8, we circle ten little ones to make a ten. There already are three tens, so in total we now have <u>four</u> tens. So, we write "4" in the tens column.

There are three little cubes left over, so we write "3" in the ones column.

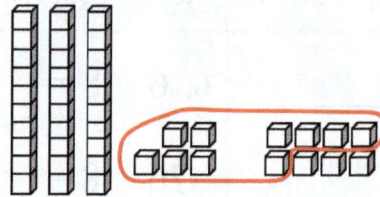

	tens	ones
	3	5
+		8
	4	3

1. **Circle** ten cubes to make **a new ten**. Count the tens, including the new one. Count the ones. Write the tens and ones in their own columns. You can also use manipulatives.

a.

	tens	ones
	3	3
+		9

b.

	tens	ones
	2	5
+		8

c.

	tens	ones
	3	8
+		9

d.

	tens	ones
	2	7
+		7

e.

	tens	ones
	3	6
+	1	8

f.

	tens	ones
	2	5
+	2	7

When we make a new ten from the ones, we are **regrouping**. The ten ones get grouped as a ten, and are counted with the other tens.

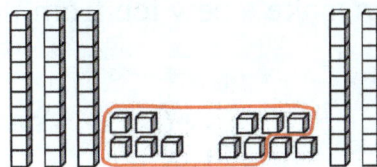

This is also called **carrying to tens**. Imagine someone "gathering" ten little cubes in his lap and "carrying" them over into the tens column as 1 ten.

	tens	ones
	1	
	3	5
+	2	7
	6	2

To show this new ten, write a little "1" in the tens column above the other numbers. Then add in the tens-column as usual, adding the little "1" also.

2. Circle ten ones to make a new ten. Add the tens and ones in columns.

a.

	tens	ones
	1	
	1	3
+	2	9
		2

b.

	tens	ones
	1	
	2	4
+	3	8

c.

	tens	ones
	1	
	3	5
+	1	9

d.

	tens	ones
	1	
	2	4
+	4	7

e.

	tens	ones
+		

f.

	tens	ones
+		

g.

	tens	ones
+		

h.

	tens	ones
+		

3. Add. If you can make a new ten from the ones, regroup.

a. 42
  + 1 5
_____

b. 27
  + 4 5
_____

c. 65
  + 2 6
_____

d. 83
  + 1 5
_____

e. 34
  + 1 9
_____

f. 52
  + 4 1
_____

g. 13
  + 4 4
_____

h. 63
  + 2 7
_____

i. 36
  + 5 1
_____

j. 66
  + 2 9
_____

We can add three numbers by writing them under each other.
This is not any more difficult than adding two numbers.

On the right, first add the ones. 2 + 7 + 5 = 14. You get a new
ten. So, regroup and write that new ten with the other tens.

In the tens, add 1 + 3 + 2 + 1 = 7.

	1	
	3	2
	2	7
+	1	5
	7	4

4. Add. Regroup the ones to make a new ten.

a. 34
    1 9
  + 2 6
_____

b. 15
    2 7
  + 4 5
_____

c. 13
    2 7
  + 2 6
_____

d. 26
    4 2
  + 1 9
_____

e. 34
    2 1
  + 1 9
_____

5. Show the additions on the number line by drawing lines that are that long.

a. 13 + 9 + 11 = _____

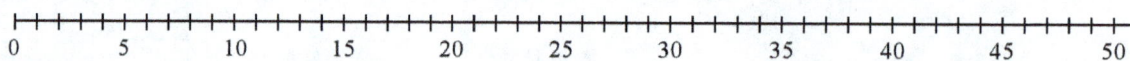

b. 27 + 16 = _____

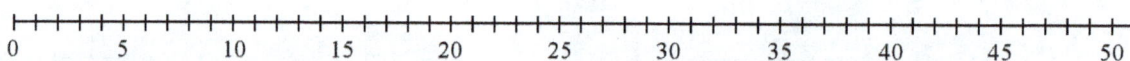

46

Add in Columns Practice

1. Add in columns.

a.
```
    9
+ 7 1
─────
```

b.
```
  2 4
+ 6 7
─────
```

c.
```
  5 5
+ 3 6
─────
```

d.
```
  4 5
+ 2 5
─────
```

e.
```
  3 8
+ 1 4
─────
```

f.
```
  3 4
    9
+ 3 5
─────
```

g.
```
  2 5
  4 2
+ 4 9
─────
```

h.
```
  5 8
  3 0
+   6
─────
```

i.
```
  2 9
  4 4
+ 1 2
─────
```

j.
```
  1 6
  1 4
+ 1 9
─────
```

2. Write the numbers so that the ones and tens are in their own columns. Add.

a. $45 + 27$
b. $8 + 56$
c. $40 + 32$
d. Double 35
e. Double 47

tens   ones   tens   ones

f. $6 + 31 + 25$
g. $40 + 7 + 9$
h. $46 + 8 + 20$
i. $5 + 8 + 13$
j. $5 + 4 + 57$

hund-reds	tens	ones
	8	6
+	6	3
1	4	9

Here we have more than 10 tens.

Ten tens make a *hundred* (100)!

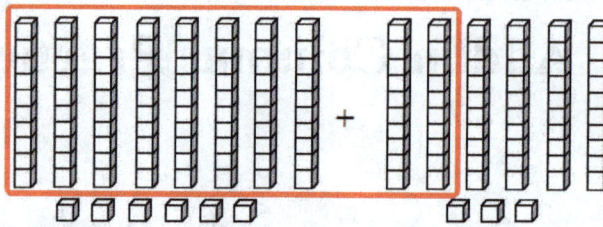

Add the tens: 8 + 6 = 14 tens. The "1" of the 14 goes in the hundreds column, and the "4" stays in the tens column. The answer 149 is read "one hundred and forty-nine."

Another example. Add the tens normally: 1 + 5 + 6 = 12 tens. Write the 12 so that the "1" is in the hundreds' column, and the "2" is in the tens column. The 12 tens make 1 hundred and 2 tens.

The answer 123 is read "one hundred and twenty-three."

You will study more about hundreds later.

hund-reds	tens	ones
	1	
	5	4
+	6	9
1	2	3

3. Add. You will have more than 10 tens.

a. 27 + 80   **b.** 95 + 47   **c.** Double 56   **d.** 62 + 84

4. Add.

a.
```
    6 7
  + 6 1
  ─────
```

b.
```
    9 0
  + 6 5
  ─────
```

c.
```
    3 9
  + 8 1
  ─────
```

d.
```
    8 5
  + 6 2
  ─────
```

e.
```
    2 9
  + 9 4
  ─────
```

f.
```
    6 5
    1 8
  + 2 6
  ─────
```

g.
```
    7 4
      7
  + 4 5
  ─────
```

h.
```
    6 8
    4 7
  + 3 2
  ─────
```

i.
```
    1 2
    8 8
  + 4 9
  ─────
```

j.
```
      8
    5 0
  + 7 9
  ─────
```

48

5. Solve the word problems. You may need to add or subtract in columns.

a. Josh worked for 27 hours this week.
Bill worked for 16 hours more than Josh.
How many hours did Bill work?

b. Natasha read 29 comic books and Matt read 16.
How many more comic books did Natasha read than Matt?

c. Mom put 13 red flowers and 11 blue flowers in one
vase. Then she put 22 flowers in another vase.
Which vase has more flowers? How many more?

d. Caleb had saved $24 and his brother David $41.
Then Caleb earned $20. Now, who has more money?
How much more?

e. Caleb bought a set of colored pencils for $13,
drawing paper for $9, and paints for $21.
What is his total bill?

Mental Addition of Two-Digit Numbers

Example 1. Add in parts 40 + 55.

First break 55 into its tens and ones. 55 is 50 + 5.

So, 40 + 55 becomes 40 + 50 + 5.

Now add 40 and 50. You get 90. Then add the 5. You get 90 + 5 = _____.

Example 2. Add in parts 36 + 30.

First break 36 into three tens and ones. 36 is 30 + 6.

So, 36 + 30 becomes 30 + 6 + 30.

Now add 30 and 30. That is 60. Then add the 6. You get 60 + 6 = _____.

1. Add *in parts*, breaking the second number into its tens and ones.

a. 20 + 34 = _____	**b.** 70 + 18 = _____	**c.** 50 + 27 = _____
20 + _____ + ____	70 + _____ + ____	50 + _____ + ____

2. Add *in parts*. Break the number that is not whole tens into its tens and ones in your mind.

a. 17 + 10 = _____	**b.** 16 + 20 = _____	**c.** 50 + 14 = _____
26 + 10 = _____	34 + 30 = _____	60 + 23 = _____
42 + 10 = _____	67 + 20 = _____	30 + 45 = _____

3. Add mentally. We already studied these. The first one is the helping problem.

a.	b.	c.	d.
7 + 8 = _____	4 + 9 = _____	8 + 4 = _____	7 + 9 = _____
17 + 8 = _____	14 + 9 = _____	48 + 4 = _____	57 + 9 = _____
37 + 8 = _____	44 + 9 = _____	78 + 4 = _____	37 + 9 = _____

How can you easily add 16 + 19? Think about it before you go on! Here's the answer: again, add *in parts*. Look at the example on the right.	$16 + 19$ $= 6 + 9 + 10 + 10$ $= 15 + 10 + 10 = $ _____

4. Add in parts.

a. $13 + 18$ $=$ ____ + ____ + 10 + 10 $=$	b. $15 + 15$ $=$ ____ + ____ + 10 + 10 $=$
c. $17 + 18$ $=$ ____ + ____ + 10 + 10 $=$	d. $19 + 15$ $=$ ____ + ____ + 10 + 10 $=$
e. $18 + 12 = $ _____	f. $13 + 16 = $ _____
g. $16 + 17 = $ _____	h. $17 + 15 = $ _____

5. **a.** Laura owns 13 cats. Five of her cats live in the house. How many of her cats live outside?

   b. Laura's cats eat 20 lb of cat food in a week. Laura has *two* 4-lb bags at home. How many more pounds of cat food does she need to have enough for one week?

6. Count by threes.

   42, 45, _____, _____, _____, _____, _____, _____, _____

7. Find the pattern and continue it. This pattern "grows" at each step.

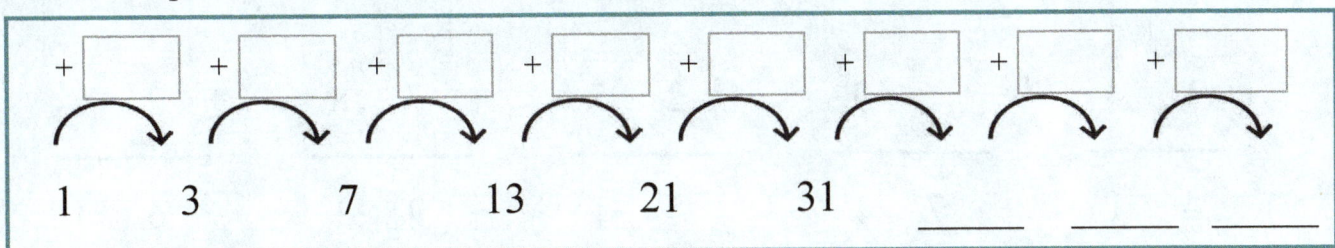

+ ☐	+ ☐	+ ☐	+ ☐	+ ☐	+ ☐	+ ☐	+ ☐

1   3   7   13   21   31   ____   ____   ____

Add two-digit numbers: Add the tens and ones separately	
Add tens on their own. Add ones on their own. Lastly, add the two sums.	$45 + 27$ $40 + 20$ + $5 + 7$ 60 + 12 = 72

8. Add by adding tens and ones separately.

a. $36 + 22$ $30 + 20$ + $6 + 2$ _____ + _____ = _____	**b.** $72 + 18$ $70 + 10$ + $2 + 8$ _____ + _____ = _____
c. $54 + 37$ $50 + 30$ + $4 + 7$ _____ + _____ = _____	**d.** $24 + 55$ ___ + ___ + ___ + ___ _____ + _____ = _____
e. $36 + 36$ ___ + ___ + ___ + ___ _____ + _____ = _____	**f.** $42 + 68$
g. $45 + 18$	**h.** $37 + 58$

Puzzle Corner Figure out the missing numbers for these addition problems.

a.	**b.**	**c.**	**d.**	**e.**
☐☐ + 1 4 ───── 4 1	☐☐ + 3 ───── 7 1	☐☐ + 2 5 ───── 5 1	☐☐ + 7 8 ───── 9 1	☐☐ + 2 6 ───── 6 1

Adding Three or Four Numbers Mentally

When you add three numbers, you can add them in any order you wish.	Perhaps add 8 and 8 first: $$8 + 8 + 6$$ $$= 16 + 6 = \underline{\hspace{1cm}}$$	Or perhaps add 8 and 6 first: $$8 + 8 + 6$$ $$= 8 + 14 = \underline{\hspace{1cm}}$$

1. Add three numbers.

a. $8 + 8 + 8 = \underline{\hspace{1.5cm}}$	b. $7 + 9 + 6 = \underline{\hspace{1.5cm}}$	c. $5 + 8 + 9 = \underline{\hspace{1.5cm}}$
d. $7 + 9 + 5 = \underline{\hspace{1.5cm}}$	e. $8 + 6 + 4 = \underline{\hspace{1.5cm}}$	f. $2 + 9 + 5 = \underline{\hspace{1.5cm}}$

When you add four numbers, often it is easy to add them *in pairs:* two numbers at a time.		But sometimes some other way of adding is easier.
Add 7 and 3. Add 5 and 6: $$7 + 5 + 3 + 6$$ $$= 10 + 11 = \underline{\hspace{1cm}}$$	Add the first two, and the last two: $$6 + 9 + 8 + 5$$ $$= 15 + 13 = \underline{\hspace{1cm}}$$	Double 8 makes 16, then to that add 4: $$9 + 8 + 8 + 4$$ $$= 16 + 4 + 9 = \underline{\hspace{1cm}}$$

2. Add four numbers. Look at the example.

a. $8 + 8 + 2 + 8$ $$= 16 + 10$$ $$= 26$$	b. $7 + 5 + 5 + 6$ $$= \underline{\hspace{1cm}} + \underline{\hspace{1cm}}$$ $$= \underline{\hspace{1cm}}$$	c. $4 + 7 + 2 + 5$ $$= \underline{\hspace{1cm}} + \underline{\hspace{1cm}}$$ $$= \underline{\hspace{1cm}}$$
d. $6 + 7 + 9 + 8$ $$= \underline{\hspace{1cm}} + \underline{\hspace{1cm}}$$ $$= \underline{\hspace{1cm}}$$	e. $8 + 5 + 2 + 6$ $$= \underline{\hspace{1cm}} + \underline{\hspace{1cm}}$$ $$= \underline{\hspace{1cm}}$$	f. $4 + 5 + 3 + 9$ $$= \underline{\hspace{1cm}} + \underline{\hspace{1cm}}$$ $$= \underline{\hspace{1cm}}$$

3. Practice adding three or four numbers.

a. $4 + 8 + 6 =$ _____	**b.** $4 + 9 + 5 + 6 =$ _____	**c.** $7 + 8 + 7 + 9 =$ _____
d. $9 + 9 + 5 =$ _____	**e.** $8 + 3 + 5 + 4 =$ _____	**f.** $2 + 6 + 6 + 5 =$ _____
g. $8 + 4 + 4 =$ _____	**h.** $9 + 2 + 4 + 6 =$ _____	**i.** $2 + 3 + 8 + 9 =$ _____

4. Madison took photos of her friends. She took eight photos
   of Mia, nine photos of Chloe, and eight of Isabella.
   How many photos did Madison take all totaled?

5. Gabriel has 7 toy cars and Lucas has 9. They put their cars together.
   Can they share the cars evenly?
   If yes, how many would each boy get?

6. Elijah made 8 sand towers and Bill made 11. Can the boys share the
   towers in a game they are playing?
   If yes, how many would each boy get?

7. Add mentally. Think, what would the *easiest order* to add the numbers!

a. $30 + 2 + 40 + 8 =$ _____	**c.** $9 + 40 + 1 + 4 =$ _____
b. $50 + 4 + 10 + 7 =$ _____	**d.** $20 + 10 + 8 + 9 =$ _____

8. Compare the expressions and write $<$, $>$, or $=$.

a. $8 + 5 + 6$ ⬚ $5 + 6 + 9$	**b.** $54 + 8$ ⬚ $53 + 9$
c. $8 + 8 + 7 + 7$ ⬚ $9 + 9 + 6 + 6$	**d.** $48 - 6$ ⬚ $38 + 5$

Adding Three or Four Numbers in Columns

Sometimes we get *two or three new* tens from the ones. We need to regroup.

In the ones, we add
$8 + 7 + 8 = 23$.

We write the two new
tens in the tens column.
Complete the problem.

```
    2
  4 8
  2 7
+ 1 8
-----
    3
```

In the ones we add $9 + 9 + 7 + 6$
$= 18 + 13 = 31$. We write *three*
new tens in the tens column.

In the tens, we add
$3 + 3 + 1 + 2 + 2 = 11$. The
answer is *more* than one hundred.
It is 111 (one hundred eleven).

```
    3
  3 9
  1 9
  2 7
+ 2 6
-----
1 1 1
```

1. Add mentally. *Remember* to first try to find if any of the numbers **make 10**.

a. $8 + 4 + 5 = $ _____

b. $3 + 8 + 7 = $ _____

c. $8 + 5 + 6 + 4 = $ _____

2. Add. The answers are "hidden" in the list of numbers below the problems.

a.
```
  5 2
  3 0
+ 1 1
-----
```

b.
```
  1 3
  2 5
+ 5 4
-----
```

c.
```
  3 3
  3 8
+ 2 7
-----
```

d.
```
  3 6
  2 7
+ 1 9
-----
```

e.
```
  3 6
  2 7
  1 8
+ 1 6
-----
```

f.
```
  4 0
  1 8
  1 6
+ 2 2
-----
```

g.
```
  1 5
  1 7
  1 8
+ 3 9
-----
```

h.
```
  1 2
  2 9
  2 5
+ 1 4
-----
```

i.
```
  1 9
  6 9
+ 1 9
-----
```

j.
```
  5 6
  3 2
+ 2 9
-----
```

k.
```
  4 5
  5 5
+ 1 9
-----
```

l.
```
  5 9
  1 9
+ 4 2
-----
```

74   80   82   89   91   92   93   96   97   98   117   107   120   119   122

55

3. Find the total bill.

a. Two shirts at $17 each; a pair of jeans for $49.

b. Three buckets for $17 each.

c. Two shovels for $24 each; two rakes for $19 each.

d. A phone for $89, a phone cover for $12, and chocolate for $7.

e. Two adult tickets for $36 each, 2 children's tickets for $23 each.

f. A child's meal for $19 and three adult meals for $29 each.

4. Find the errors in these additions, and correct them.

a.
```
   3 3
 + 4 8
 ─────
   711
```

b.
```
   5 5
 + 3 9
 ─────
   814
```

5. Solve the problems. You need to add or subtract.

a. One bus has 35 people on it, and another has 22.
How many more people does the first one have than the second?

b. A bus had some people in it. Then, 13 more people
got on. Now there are 19 people on the bus.
How many were on the bus originally?

c. One bus can seat 40 people. There were already 33 people.
Is there room for nine more people?

Yes/No, because

d. One bus can seat 40 people.
How many buses do you need for 76 people?

How many buses do you need for 99 people?

e. A bus was full with 40 people, but then six people got off.
How many people are on the bus now?

f. A bus was full with 40 people. First it dropped off 3 people.
Then it dropped off seven more people. How many people
were left on the bus?

6. Add.

a.	b.	c.	d.
3 9	3 3	1 7	5 5
1 5	4 8	3 7	1 8
1 8	1 6	2 5	1 5
+ 2 8	+ 1 3	+ 3 4	+ 2 7

7. Are these numbers even or odd? Mark an "X". If the number is even, write it as a double of some number.

Number	Even?	Odd?	As a double:
8	X		4 + 4
16			
100			
19			

Number	Even?	Odd?	As a double:
18			
24			
15			
21			

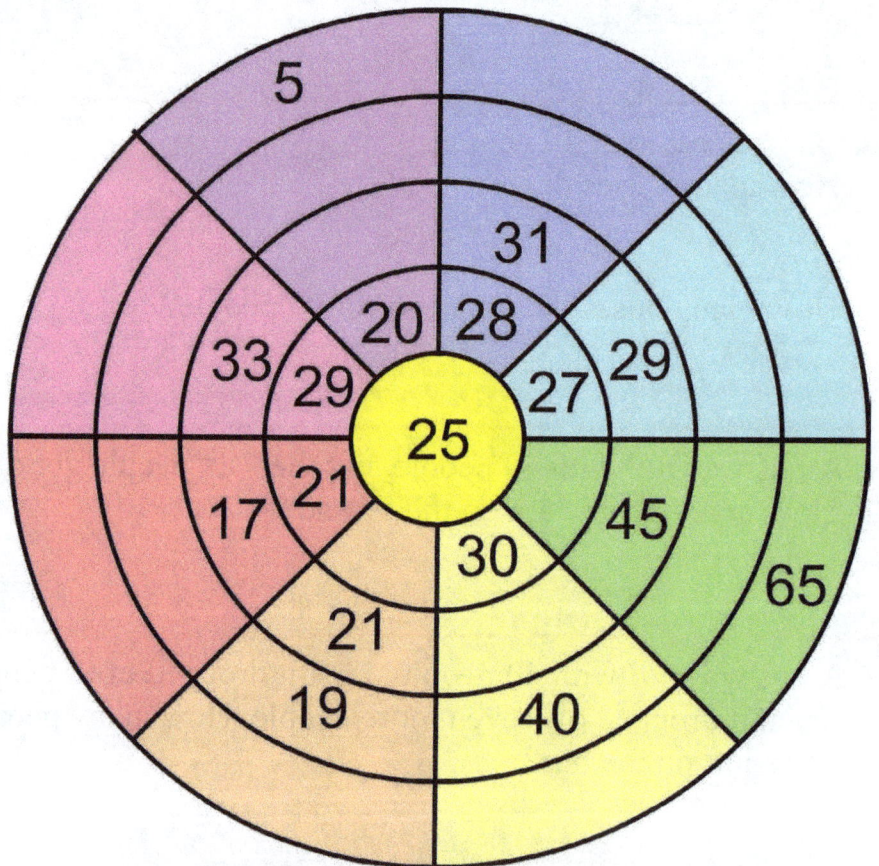

Puzzle Corner

Skip-count from 25 (in the middle) to the outer edge. Each sector has a different skip-counting pattern— either by 2s, by 3s, by 4s, by 5s, or by 10s.

Regrouping in Subtraction, Part 1

We will now study regrouping (also called "borrowing") in subtraction.

As a first step, we break a ten-pillar into ten little cubes. This is called *regrouping*, because one ten "changes groups" from the tens group into the ones.

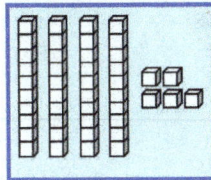

4 tens 5 ones

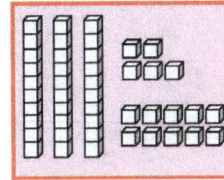

3 tens 15 ones

First we have 45. We "break" one ten-pillar into little cubes.

Now we have 3 tens and 15 ones. It is still 45, but written in a different way.

Here is another example. First we have 5 tens 3 ones. We "break" one ten-pillar into 10 little cubes. We end up with 4 tens 13 ones.

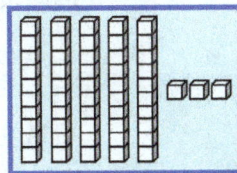

5 tens 3 ones

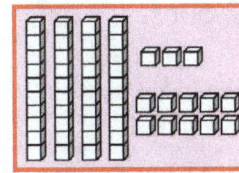

4 tens 13 ones

1. Break a ten into 10 ones. What do you get? Draw or use manipulatives to help.

a.
3 tens   0 ones ➡ ___ tens ___ ones

b.
___ tens ___ ones ➡ ___ tens ___ ones

c.
___ tens ___ ones ➡ ___ tens ___ ones

d.
___ tens ___ ones ➡ ___ tens ___ ones

e.
___ tens ___ ones ➡ ___ tens ___ ones

f.
___ tens ___ ones ➡ ___ tens ___ ones

Let's study subtraction. The pictures on the right illustrate 45 – 17.

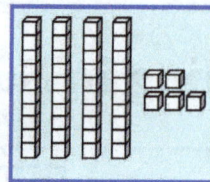 Break a ten.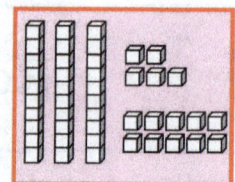

4 tens 5 ones   3 tens 15 ones

First, a ten is broken into 10 ones. So, 4 tens 5 ones becomes 3 tens 15 ones.

After that, cross out (subtract) 1 ten 7 ones.

Cross out 1 ten 7 ones (from the *second* picture).

What is left? _____ tens _____ ones

The pictures on the right illustrate 52 – 39.

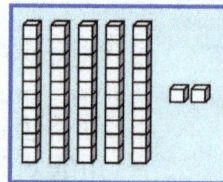 Break a ten.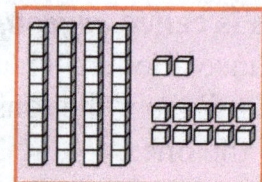

5 tens 2 ones   4 tens 12 ones

First, a ten is broken into 10 ones. So, 5 tens 2 ones becomes 4 tens 12 ones.

After that, cross out (subtract) 3 tens 9 ones.

Cross out 3 tens 9 ones (from the *second* picture).

What is left? _____ tens _____ ones

---

2. Fill in. Always subtract (cross out some) from the *second* picture.

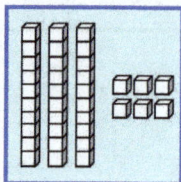 Break a ten.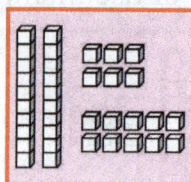

3 tens   6 ones   2 tens   16 ones

a. Subtract 8 ones (from the *second* picture).

What is left? _____ tens _____ ones

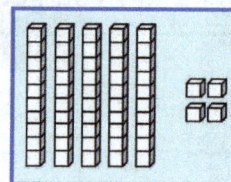 Break a ten.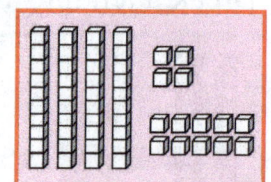

___ tens ___ ones   ___ tens ___ ones

b. Subtract 2 tens 7 ones.

What is left? _____ tens _____ ones

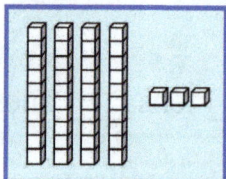 Break a ten.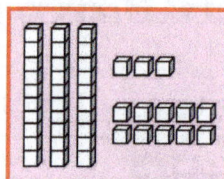

___ tens ___ ones   ___ tens ___ ones

c. Cross out 2 tens 5 ones.

What is left? _____ tens _____ ones

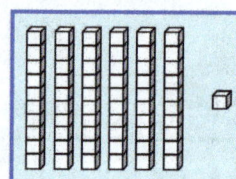 Break a ten.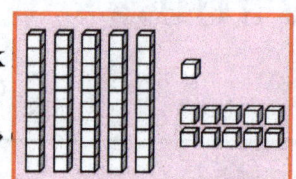

___ tens ___ ones   ___ tens ___ ones

d. Cross out 4 tens 4 ones.

What is left? _____ tens _____ ones

3. First, break a ten. Then subtract ones and tens separately. Look at the example.

a.
 	
5 tens 5 ones ⟹   4 tens   15 ones
             −   3   tens   7   ones
                     1   ten   8   ones
	b.
	
	7 tens 2 ones ⟹ ___ tens ___ ones
	             − 3 tens 5 ones
	              ___ tens ___ ones

c.
 	
6 tens 0 ones ⟹ ___ tens ___ ones
             −   2   tens   7   ones
              ___ tens ___ ones
	d.
	
	6 tens 4 ones ⟹ ___ tens ___ ones
	             − 3 tens 8 ones
	              ___ tens ___ ones

e.
 	
7 tens 6 ones ⟹ ___ tens ___ ones
             −   4   tens   7   ones
              ___ tens ___ ones
	f.
	
	5 tens 0 ones ⟹ ___ tens ___ ones
	             − 2 tens 2 ones
	              ___ tens ___ ones

g.
 	
8 tens 1 one ⟹ ___ tens ___ ones
             −   6   tens   5   ones
              ___ tens ___ ones
	h.
	
	6 tens 3 ones ⟹ ___ tens ___ ones
	             − 2 tens 8 ones
	              ___ tens ___ ones

4. Jessica had 27 colored pencils and her brother and sister had none. Then Jessica gave 10 of them to her brother, and four to her sister.

   a. How many pencils does Jessica have now?

   b. How many more pencils does Jessica have than her brother?

   c. How many more pencils does Jessica have than her sister?

Regrouping in Subtraction, Part 2

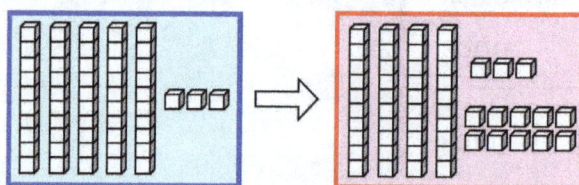

5 tens, 3 ones   4 tens, 13 ones

Cross out 1 ten 6 ones. What is left? _____ tens _____ ones

In columns:

	4	13
	~~5~~	~~3~~
−	1	6
	3	7

When the subtraction is done *in columns*:

- We take (borrow) one ten from the 5 tens. There will be now only 4 tens in the tens column, so to show this, we **cross the "5"** in the tens column and **write 4 above it**.

- This new ten is now grouped with the ones. There were 3 ones, but with the 10 new ones there will be 13. To show this, we also **cross the "3"** in the ones column and **write 13 above it**.

- Then we subtract the tens and ones separately.

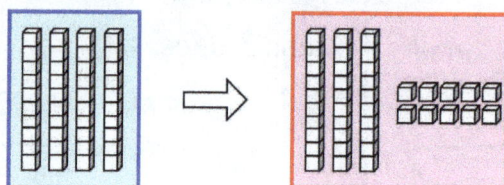

4 tens   3 tens, 10 ones

Cross out 2 tens 8 ones. What is left? ____ tens ____ ones

	3	10
	4	~~0~~
−	2	8
	1	2

Here is another example: 40 − 28.

- We take (borrow) one ten from the 4 tens. There will be now only 3 tens in the tens column, so to show this, we **cross the "4"** in the tens column and **write 3 above it**.

- This new ten is now grouped with the ones. There were 0 ones, but with the 10 new ones there will be 10. To show this, we also **cross the "0"** in the ones column and **write 10 above it**.

- Then we subtract the tens and ones separately in columns.

1. Regroup first. Then subtract.

a. 6 tens 0 ones → _____ tens _____ ones

Take away
3 tens, 9 ones.

	6	0
−	3	9

b. 7 tens 1 one → _____ tens _____ ones

Take away
1 ten, 6 ones.

	7	1
−	1	6

c. 3 tens, 5 ones → _____ tens _____ ones

Take away
1 ten, 7 ones.

	3	5
−	1	7

d. 8 tens → _____ tens _____ ones

Take away
3 tens, 4 ones.

	8	0
−	3	4

e. 7 tens, 6 ones → _____ tens _____ ones

Take away
4 tens, 8 ones.

	7	6
−	4	8

f. 9 tens → _____ tens _____ ones

Take away
5 tens, 1 one.

	9	0
−	5	1

g. 5 tens, 4 ones → _____ tens _____ ones

Take away
2 tens, 5 ones.

−		

h. 8 tens → _____ tens _____ ones

Take away
4 tens, 7 ones.

−		

i. 7 tens, 4 ones → _____ tens _____ ones

Take away
3 tens, 8 ones.

−		

j. 4 tens 7 ones → _____ tens _____ ones

Take away
2 tens, 9 ones.

−		

2. Subtract. Check by adding the result and what was subtracted.

a.	Check:	b.	Check:	c.	Check:
4 16 $\cancel{5}\;\cancel{6}$ $-\;2\;7$ _____ 2 9	1 2 9 $+\;2\;7$ _____ 5 6	9 0 $-\;2\;8$ _____	$+\;2\;8$ _____	4 2 $-\;1\;5$ _____	$+\;1\;5$ _____
d. 9 0 $-\;3\;5$ _____		e. 8 2 $-\;2\;5$ _____		f. 6 5 $-\;3\;9$ _____	
g. 5 2 $-\;1\;4$ _____		h. 6 5 $-\;2\;6$ _____		i. 7 0 $-\;4\;8$ _____	
j. 5 5 $-\;1\;7$ _____		k. 3 1 $-\;1\;8$ _____		l. 6 6 $-\;2\;8$ _____	

Figure out the missing numbers in these subtractions! **Puzzle Corner**
You might need to regroup.

$$\begin{array}{r} \square\;3 \\ -\;1\;\square \\ \hline 7\;5 \end{array} \qquad \begin{array}{r} 8\;\square \\ -\;\square\;7 \\ \hline 1\;6 \end{array} \qquad \begin{array}{r} \square\;0 \\ -\;3\;\square \\ \hline 4\;2 \end{array} \qquad \begin{array}{r} \square\;\square \\ -\;1\;4 \\ \hline 6\;8 \end{array} \qquad \begin{array}{r} 6\;2 \\ -\;\square\;\square \\ \hline 5\;3 \end{array}$$

64

Regrouping in Subtraction, Part 3

Sometimes we need to regroup in subtraction, and sometimes not. Check carefully in the ones column. Are there enough ones to do the subtraction, or not? If not, you need to regroup.	Do you need to regroup? YES / NO $\begin{array}{r} 6\ 1 \\ -\ 2\ 6 \\ \hline \end{array}$	Do you need to regroup? YES / NO $\begin{array}{r} 7\ 4 \\ -\ 2\ 3 \\ \hline \end{array}$

1. Look at the ones' digits. Do you need to regroup (borrow a ten in the ones' column)?

a. Do you need to regroup? YES / NO $\begin{array}{r} 5\ 4 \\ -\ 3\ 2 \\ \hline \end{array}$	**b.** Do you need to regroup? YES / NO $\begin{array}{r} 5\ 0 \\ -\ 2\ 5 \\ \hline \end{array}$	**c.** Do you need to regroup? YES / NO $\begin{array}{r} 8\ 2 \\ -\ 5\ 6 \\ \hline \end{array}$

2. Subtract. Regroup if necessary. Find the answers in the line of numbers below.

a. Do you need to regroup? YES / NO $\begin{array}{r} 6\ 0 \\ -\ 1\ 6 \\ \hline \end{array}$	**b.** Do you need to regroup? YES / NO $\begin{array}{r} 5\ 7 \\ -\ 3\ 2 \\ \hline \end{array}$	**c.** Do you need to regroup? YES / NO $\begin{array}{r} 4\ 3 \\ -\ 1\ 7 \\ \hline \end{array}$

d. $\begin{array}{r} 8\ 0 \\ -\ 2\ 8 \\ \hline \end{array}$	**e.** $\begin{array}{r} 9\ 7 \\ -\ 2\ 5 \\ \hline \end{array}$	**f.** $\begin{array}{r} 8\ 1 \\ -\ 5\ 7 \\ \hline \end{array}$	**g.** $\begin{array}{r} 3\ 7 \\ -\ 2\ 7 \\ \hline \end{array}$	**h.** $\begin{array}{r} 6\ 0 \\ -\ 4\ 1 \\ \hline \end{array}$

44   26   19   25   24   72   10   52   25

3. Subtract mentally.

a. $64 - 20 =$ _____ $98 - 50 =$ _____	**b.** $77 - 71 =$ _____ $45 - 40 =$ _____	**c.** $98 - 6 =$ _____ $50 - 46 =$ _____

The number line arrows illustrate the subtraction 23 − 7. The first, red, arrow goes from 0 to 23. The second, blue, arrow goes 7 steps backwards from 23 and ends at 16.

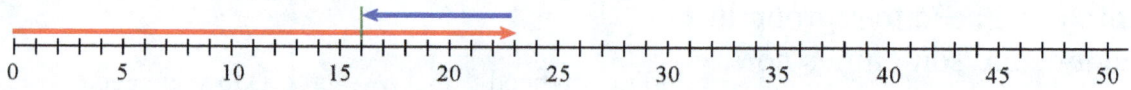

4. Write the subtractions illustrated by the arrows on the number line.

a.

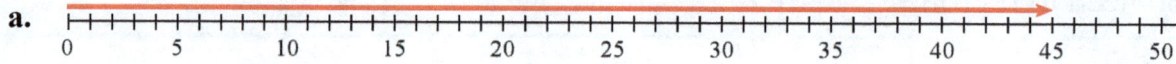

b.

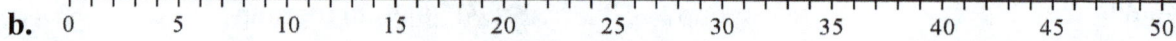

c.

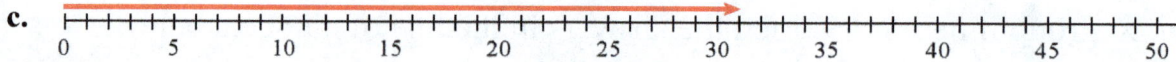

5. Draw arrows to illustrate these subtractions on the number line.

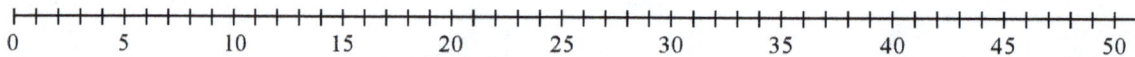

a. $22 - 9 = $ _____

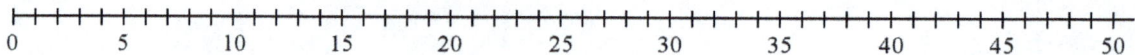

b. $36 - 12 = $ _____

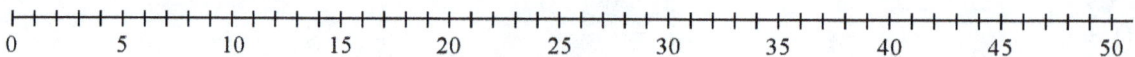

c. $44 - 17 = $ _____

Remember, subtraction and addition are
connected. For example, $9 - 4 = 5$ and
$5 + 4 = 9$.

You can use this connection, and <u>check</u> each
subtraction by adding.

Add the answer you got *and* the number you
subtracted. You should get the number you
subtracted from.

For example, to check if $68 - 45$ is really 23,
add $23 + 45$ and check if you get 68.

$$
\begin{array}{r} {}^{4}\;{}^{14} \\ \cancel{5}\;\cancel{4} \\ -\;3\;7 \\ \hline 1\;7 \end{array}
\qquad
\begin{array}{r} {}^{1} \\ 1\;7 \\ +\;3\;7 \\ \hline 5\;4 \end{array}
$$

6. Subtract. Regroup if necessary. Check each subtraction by *adding your answer
and the number you subtracted.*

a.

$$
\begin{array}{r} {}^{8}\;{}^{14} \\ \cancel{9}\;4 \\ -\;3\;5 \\ \hline 5\;9 \end{array}
\qquad
\begin{array}{r} 5\;9 \\ +\;3\;5 \\ \hline \end{array}
$$

b.

$$
\begin{array}{r} 8\;2 \\ -\;2\;5 \\ \hline \end{array}
\qquad
\begin{array}{r} \\ +\;2\;5 \\ \hline \end{array}
$$

c.

$$
\begin{array}{r} 6\;1 \\ -\;4\;9 \\ \hline \end{array}
\qquad
\begin{array}{r} \\ +\;4\;9 \\ \hline \end{array}
$$

d.

$$
\begin{array}{r} 9\;9 \\ -\;5\;7 \\ \hline \end{array}
\qquad
\begin{array}{r} \\ +\;5\;7 \\ \hline \end{array}
$$

e.

$$
\begin{array}{r} 6\;0 \\ -\;2\;3 \\ \hline \end{array}
\qquad
\begin{array}{r} \\ +\;2\;3 \\ \hline \end{array}
$$

f.

$$
\begin{array}{r} 6\;6 \\ -\;4\;8 \\ \hline \end{array}
\qquad
\begin{array}{r} \\ +\;4\;8 \\ \hline \end{array}
$$

g.

$$
\begin{array}{r} 5\;4 \\ -\;4\;1 \\ \hline \end{array}
\qquad
\begin{array}{r} \\ + \\ \hline \end{array}
$$

h.

$$
\begin{array}{r} 8\;5 \\ -\;3\;9 \\ \hline \end{array}
\qquad
\begin{array}{r} \\ + \\ \hline \end{array}
$$

7. Solve. IF you subtract, check the answer by adding (you will not subtract in every problem).

a. Emily picked 29 rows of strawberries and Jim picked 13 rows of strawberries. How many more rows of strawberries did Emily pick?

b. Judith sold 35 tickets for the county fair and Peter sold 62 tickets. How many more tickets did Peter sell than Judith?

c. Judith sold 35 tickets for the county fair and Peter sold 62 tickets. How many tickets did they sell together?

d. A pretty doll with a blue dress costs $28, and a different doll with a pink dress costs $12 more than that. How much does that doll cost?

e. Bill bought two bicycle chains for $18 each and a saddle for $49. How much was the total cost?

Word Problems

"More" in word problems

Study carefully these problems. They all use the word MORE, but they are different! Solve each problem with your teacher or on your own, if you can. In each problem **think** first, "WHO has more?" (If the problems are difficult, drawing the situations may help also.)

* Anna has 12 sheep. Her neighbor has 7 more sheep than Anna.
  How many sheep does the neighbor have?

* Anna has 7 more sheep than her neighbor. Anna has 12 sheep.
  How many sheep does the neighbor have?

* Anna has 12 sheep, and her neighbor has 7.
  How many more sheep does Anna have than her neighbor?

1. Solve. Think if you need addition or subtraction.

a. Isabella has a flock of 15 goats. Her neighbor Andy has 18 goats.
How many more goats does Andy have than Isabella?

b. Isabella has a flock of 15 goats. Her neighbor Sandy has 8 more goats than Isabella.
How many goats does Sandy have?

c. Isabella has 15 goats, which is 5 more than what Henry has.
How many goats does Henry have?

d. Christopher has 27 cows, and Daniel has 16 more than that.
How many cows does Daniel have?

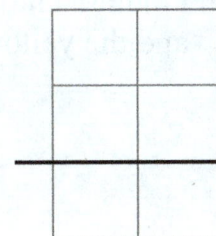

"Fewer" (or "less") in word problems

FEWER is the opposite of MORE. These three problems all use the word FEWER, but they are different! Solve each problem with your teacher or on your own, if you can. In each problem **think** first, "WHO has more?"

- Anna has 12 sheep. Her neighbor has 7 fewer sheep than Anna. How many sheep does the neighbor have?

- Anna has 7 fewer sheep than her neighbor. Anna has 12 sheep. How many sheep does the neighbor have?

- Anna has 7 sheep, and her neighbor has 12. How many fewer sheep does Anna have than her neighbor?

2. Solve. Think if you need addition or subtraction.

a. Joe has 27 tennis balls and Mason has 5 fewer tennis balls than Joe. How many tennis balls does Mason have?

b. Joe has 27 tennis balls, which is 7 less than what Logan has. How many tennis balls does Logan have?

c. Liz wants to buy a blue dress that costs $41. A white dress, costs $13 less than that. And a yellow dress costs $3 less than the white dress.

How much does the yellow dress cost?

d. Find how much it is if Liz buys both the blue and the yellow dress.

3. Subtract mentally.

a. $30 - 28 =$ _____	**d.** $70 - 63 =$ _____	**g.** $56 - 5 =$ _____
b. $52 - 30 =$ _____	**e.** $70 - 6 =$ _____	**h.** $18 - 7 =$ _____
c. $98 - 90 =$ _____	**f.** $100 - 5 =$ _____	**i.** $46 - 4 =$ _____

4. Solve. You may need to both add and subtract.

a. Ryan rode his horse for 11 km each day for two days. Zoe rode her horse for 8 km each day for two days.

In two days, how many kilometers less did Zoe ride hers than what Ryan rode his?

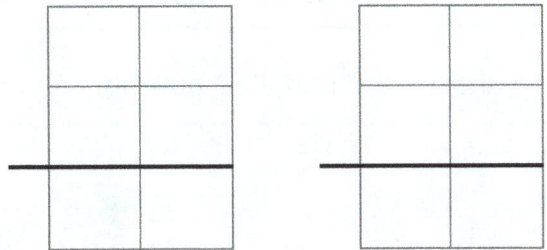

b. Mia owns 32 dolls and her friend Ava has less. Actually, Ava has 8 fewer dolls than Mia. How many dolls do the girls have in total?

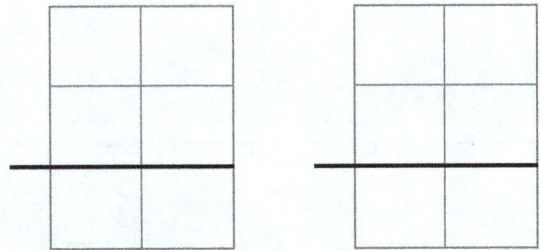

Puzzle Corner Find out what number the triangle means. You are solving *equations*!

a. $63 + \triangle = 71$

$\triangle =$ _____

b. $80 - \triangle = 54$

$\triangle =$ _____

c. $\triangle - 10 = 40$

$\triangle =$ _____

Write your own "triangle problems" (equations), and let a friend solve them.

d.

$\triangle =$ _____

e.

$\triangle =$ _____

f.

$\triangle =$ _____

Mental Subtraction, Part 1

Method 1:   Subtract in two parts

$53 - \boxed{8}$

$= 53 - \boxed{3} - \boxed{5}$

$= \quad 50 \quad - 5 = 45$

$72 - \boxed{6}$

$= 72 - \boxed{2} - \boxed{4}$

$= \quad 70 \quad - 4 = 66$

Subtract 8 in two parts: first 3, then 5.   Subtract 6 in two parts: first 2, then 4.

In other words, first subtract to the *previous whole ten*, then the rest.

1. Subtract the elevated number in parts.

-5 a. $51 - \boxed{1} - \boxed{4} =$ _____	-7 b. $62 -$ ____ $-$ ____ $=$ _____
-4 c. $33 -$ ____ $-$ ____ $=$ _____	-5 d. $92 -$ ____ $-$ ____ $=$ _____
-6 e. $75 -$ ____ $-$ ____ $=$ _____	-7 f. $63 -$ ____ $-$ ____ $=$ _____
-7 g. $35 -$ ____ $-$ ____ $=$ _____	-5 h. $74 -$ ____ $-$ ____ $=$ _____

2. First subtract the balls that are not in the ten-groups.

a. $51 - 7 =$ _____ $51 - 5 =$ _____ $51 - 3 =$ _____ $51 - 6 =$ _____	b. $42 - 4 =$ _____ $42 - 5 =$ _____ $42 - 3 =$ _____ $42 - 6 =$ _____

> ## Method 2:   Use known subtraction facts
>
> Since $14 - 6 = 8$, we know that the answer to $74 - 6$ will end in 8, but it will be sixty-something. So it is 68.
>
> Since $15 - 8 = 7$, we know that the answer to $55 - 8$ will end in 7, but it will be forty-something. So it is 47.

3. Subtract. The first problem in each box is a "helping problem" for the others.

a. $14 - 9 =$ _____ $24 - 9 =$ _____ $44 - 9 =$ _____	**b.** $17 - 8 =$ _____ $27 - 8 =$ _____ $37 - 8 =$ _____	**c.** $12 - 9 =$ _____ $52 - 9 =$ _____ $32 - 9 =$ _____
d. $15 - 9 =$ _____ $65 - 9 =$ _____ $45 - 9 =$ _____	**e.** $13 - 8 =$ _____ $33 - 8 =$ _____ $93 - 8 =$ _____	**f.** $16 - 8 =$ _____ $86 - 8 =$ _____ $36 - 8 =$ _____

4. **a.** Amy has $32. She bought a comic book for $7.
   How much does she have now?

   b. Peter had $29. A toy train he wants costs $39.
   Mom paid him $5 for working. How much more
   does Peter now need to buy the train?

   c. A flower shop has 55 roses. Eight of them are white,
   and the rest are red. How many are red?

5. Use either method from this lesson to subtract.

a.	b.	c.	d.
$34 - 5 =$ _____ $73 - 7 =$ _____	$65 - 9 =$ _____ $36 - 8 =$ _____	$51 - 8 =$ _____ $93 - 6 =$ _____	$62 - 7 =$ _____ $83 - 8 =$ _____

Mental Subtraction, Part 2

Method 3:   Add.

You can "add backwards". This works well if the two numbers are close to each other.

Instead of subtracting, think how much you need to add to the number being subtracted (the subtrahend) in order to get the number you are subtracting from (the minuend).

Think: 84 + ▭ = 92

(84 and how many more makes 92?)

$92 - 84 =$ _____

Think: 25 + ▭ = 75

(25 and how many more makes 75?)

$75 - 25 =$ _____

1. To find these differences, think of adding more.

+ 8	+	+
a. $92 - 84 =$ _____	b. $51 - 49 =$ _____	c. $76 - 69 =$ _____
(Think: 84 + ____ = 92)	(Think: 49 + ___ = 51)	(Think: 69 + ___ = 76)
d. $32 - 28 =$ _____	g. $90 - 83 =$ _____	j. $100 - 95 =$ _____
e. $22 - 14 =$ _____	h. $64 - 56 =$ _____	k. $64 - 55 =$ _____
f. $53 - 46 =$ _____	i. $72 - 65 =$ _____	l. $44 - 37 =$ _____

2. Solve. Think if you need addition or subtraction—or both.

a. Jerry has 46 toy cars. Larry has 7 more than Jerry, and Mickey has 7 less than Jerry. How many toy cars does Larry have?

And Mickey?

b. Andy has $33. He bought a gift for his mom that cost $28. Then, Andy got $5 from his dad for helping with car repairs. How much money does Andy have now?

Method 4:	Subtract in parts: tens and ones

Break the number being subtracted into its tens and ones. Subtract in parts.

$$53 - \boxed{21}$$
$$= 53 - \boxed{20} - \boxed{1}$$
$$= 33 \quad - 1 = 32$$

First subtract 20, then 1.

$$87 - \boxed{46}$$
$$= 87 - \boxed{40} - \boxed{6}$$
$$= 47 \quad - 6 = 41$$

First subtract 40, then 6.

3. Solve.

a. $78 - 22$	b. $56 - 31$	c. $46 - 25$
THINK: $78 - 20 - 2$	THINK: $56 - 30 - 1$	$46 - \underline{\quad} - \underline{\quad}$
$= \underline{\quad}$	$= \underline{\quad}$	$= \underline{\quad}$
d. $66 - 43$	e. $28 - 12$	f. $84 - 52$
$66 - \underline{\quad} - \underline{\quad}$	$28 - \underline{\quad} - \underline{\quad}$	$84 - \underline{\quad} - \underline{\quad}$
$= \underline{\quad}$	$= \underline{\quad}$	$= \underline{\quad}$
g. $99 - 66 = \underline{\quad}$	h. $47 - 34 = \underline{\quad}$	i. $78 - 67 = \underline{\quad}$

4. a. Noah counted books on his bookshelf. One shelf had 34 books. Another had 42. How many more books did that shelf have than the first?

   b. Noah took four books from the second shelf and put them on the first one. Now how many more books does the second shelf have than the first?

5. Draw arrows to illustrate the subtraction on the number line.

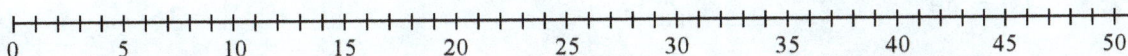

$35 - 18 = \underline{\quad}$

Method 5: Subtract in parts: tens and ones

Break BOTH the number you subtract from AND the number being subtracted into its tens and ones. Subtract the tens. Subtract the ones.

$53 \quad - \quad 21$	$76 \quad - \quad 33$
$= 50 - 20$ and $3 - 1$	$= 70 - 30$ and $6 - 3$
$= \quad 30$ and $2 = 32$	$= \quad 40$ and $3 = 43$

6. Solve.

a. $67 - 53$	b. $92 - 31$	c. $85 - 22$
$60 - 50$ and $7 - 3$ $= \underline{\quad}$	$90 - 30$ and $2 - 1$ $= \underline{\quad}$	$80 - 20$ and $\underline{\quad} - \underline{\quad}$ $= \underline{\quad}$
d. $88 - 62 = \underline{\quad}$	e. $57 - 23 = \underline{\quad}$	f. $79 - 17 = \underline{\quad}$

7. **a.** Terry is on page 52 of her book. The book has a total of 95 pages. How many pages does she have left to read?

   b. Terry reads nine pages more. Now how many pages does she have left to read?

8. Devise your own method for these subtractions. Explain how your method works.

a. $52 - 36$	b. $81 - 47$

Puzzle Corner The triangle and square represent "mystery numbers." Find them. Guess and check! Then change your guess.

a.

$\triangle + \triangle + 10 = 34$

$\triangle = \underline{\quad}$

b. $\square + \triangle = 22$

$\square - \triangle = 4$

$\triangle = \underline{\quad}, \square = \underline{\quad}$

c. $\square + \triangle = 22$

$\square + \square = 36$

$\triangle = \underline{\quad}, \square = \underline{\quad}$

Review

1. Add *in parts*. Break the number that is not whole tens into its tens and ones in your mind.

a. $17 + 10 =$ _____ $42 + 10 =$ _____	**b.** $16 + 20 =$ _____ $67 + 20 =$ _____	**c.** $50 + 14 =$ _____ $30 + 45 =$ _____

2. Add.

a. $27 + 8 =$ _____ $54 + 7 =$ _____	**b.** $18 + 9 =$ _____ $73 + 8 =$ _____	**c.** $5 + 87 =$ _____ $7 + 88 =$ _____

3. Add by adding tens and ones separately.

a.　　　$36 + 22$ $30 + 20 + 6 + 2$ _____ + _____ = _____	**b.**　　　$72 + 18$ $70 + 10 + 2 + 8$ _____ + _____ = _____
c.　　　$54 + 37$ $50 + 30 + 4 + 7$ _____ + _____ = _____	**d.**　　　$24 + 55$ _____ + _____ + _____ + _____ _____ + _____ = _____

4. Solve the problems.

a. Diane and Ted picked fruit for Mr. Mohan. Diane earned $25 and Ted earned double that. How much did Ted earn?

How much did the two earn together?

b. Emily has 24 flower plants in her yard. Leah has half that many. How many flower plants does Leah have?

5. Add.

a.
```
  4 3
+ 2 8
-----
```

b.
```
  3 3
+ 3 9
-----
```

c.
```
  2 4
+ 4 7
-----
```

d.
```
  2 3
+ 3 8
-----
```

e.
```
  5 5
+ 1 7
-----
```

f.
```
  3 8
  1 3
+ 4 2
-----
```

g.
```
  3 9
  1 0
+ 4 6
-----
```

h.
```
  4 1
  4 4
+ 3 6
-----
```

i.
```
  3 8
    7
  4 9
+ 2 3
-----
```

j.
```
  2 7
  3 6
  1 9
+ 3 5
-----
```

6. Solve.

a. Naomi bought some potatoes for $18, onions for $15, and meat for $40. What was the total cost?

b. If you buy three chairs for $34 each, what is the total bill?

c. Anna has 29 stickers and so does Betty. Ruth has 22 and Judy has 26. How many stickers are there total?

d. Andy had $47 in his wallet. He earned $15 by selling lemonade. Now can he buy a remote-controlled toy car for $65?

If yes, how much money would he have left after buying it?

If not, how much more money would he need?

78

Answer Key

Adding Within the Same Ten, p. 9

1.

a. 31 + 3 = 34	b. 32 + 6 = 38
c. 43 + 3 = 46	d. 47 + 2 = 49

2.

a. 5 + 2 = 7 35 + 2 = 37	b. 4 + 5 = 9 64 + 5 = 69	c. 3 + 6 = 9 93 + 6 = 99

3.

a. 52 + 7 = 59 2 + 7 = 9	b. 33 + 1 = 34 3 + 1 = 4	c. 11 + 5 = 16 1 + 5 = 6

4.

a. 35 + 3	b. 12 + 6	c. 57 + 1	d. 64 + 3
ones 3 | 5 + ↓ | 3 3 | 8	tens ones 1 | 2 + ↓ | 6 1 | 8	tens ones 5 | 7 + ↓ | 1 5 | 8	tens ones 6 | 4 + ↓ | 3 6 | 7

5.

a. 26 + 3	b. 72 + 4	c. 65 + 4	d. 81 + 4
tens ones 2 | 6 + ↓ | 3 2 | 9	tens ones 7 | 2 + ↓ | 4 7 | 6	tens ones 6 | 5 + ↓ | 4 6 | 9	tens ones 8 | 1 + ↓ | 4 8 | 5

6.

a. 6 + 2 = 8 16 + 2 = 18 36 + 2 = 38	b. 4 + 3 = 7 24 + 3 = 27 34 + 3 = 37	c. 5 + 4 = 9 45 + 4 = 49 65 + 4 = 69	d. 11 + 7 = 18 61 + 7 = 68 41 + 7 = 48

7.

a. 20 + 5 + 2 = 27 44 + 2 + 2 = 48	b. 93 + 1 + 5 = 99 83 + 4 + 3 = 90	c. 100 + 4 + 5 = 109 52 + 4 + 2 = 58

8.

a.	b.	c.
$18 = 10 + 8$ $25 = 20 + 5$ $55 = 50 + 5$	$32 = 30 + 2$ $95 = 90 + 5$ $49 = 40 + 9$	$66 = 60 + 6$ $89 = 9 + 80$ $78 = 8 + 70$

9.

a. $24 + 3$ $<$ $24 + 5$	c. $17 + 2$ $<$ $19 + 2$	e. 58 $<$ $8 + 51$
b. $83 + 5$ $=$ $85 + 3$	d. $36 + 4$ $<$ $46 + 4$	f. 66 $=$ $5 + 61$

Puzzle Corner. The one marked with ? is the comparison you cannot do without knowing the mystery number. Why is that? Because the comparison depends on the value of the star. If ⭐ is a large number, such as 100, then ⭐ + ⭐ > ⭐ + 20. But if ⭐ is a small number such as 2, then ⭐ + ⭐ < ⭐ + 20.

⭐ + 5 $>$ ⭐ + 4   ⭐ − 5 $<$ ⭐ − 4   ⭐ − 5 $<$ ⭐

⭐ + 2 $<$ ⭐ + 7   ⭐ − 5 $>$ ⭐ − 6   ⭐ + ⭐ $?$ ⭐ + 20

Subtracting Within the Same Ten, p. 12

1. a. 6, 26   b. 1, 11   c. 0, 60   d. 0, 50   e. 1, 41   f. 3, 93

2. a. 52, 2   b. 74, $6 - 2 = 4$ c. 84, $8 - 4 = 4$

3. a. 31, 32, 33   b. 50, 52, 54   c. 46, 44, 42   d. 33, 32, 30

4. a. 71, 21   b. 45, 74   c. 53, 84   d. 12, 94

5. a. 5, 6, 7   b. 1, 1, 4   c. 4, 3, 7

6. a. She sold 28 pictures in all. $21 + 7 = 28$
   b. She has two left. $28 + 2 = 30$ or $30 - 28 = 2$
   c. At 7:30. She took three hours, and three hours later than 4:30 is 7:30.

7. a. $37 - 7 = 30$ b. $46 - 6 = 40$ c. $28 - 8 = 20$
   d. $57 - 7 = 50$ e. $85 - 5 = 80$ f. $69 - 9 = 60$

8.

a. $50 + \underline{7} = 57$	b. $\underline{86} + 2 = 88$	c. $79 - 9 = \underline{70}$
d. $\underline{25} - 5 = 20$	e. $90 - \underline{5} = 85$	f. $42 = 40 + \underline{2}$

9. a. 10, 15, 20, 25, 30, 35, 40, 45, 50
   b. 1, 6, 11, 16, 21, 26, 31, 36, 41
   c. 3, 8, 13, 18, 23, 28, 33, 38, 43

10.

a.	b.	c.
88 − 0 = <u>88</u>	95 − 2 = <u>93</u>	48 − 1 = <u>47</u>
88 − 1 = <u>87</u>	85 − 2 = <u>83</u>	46 − 1 = <u>45</u>
88 − 2 = <u>86</u>	75 − 2 = <u>73</u>	44 − 1 = <u>43</u>
88 − <u>3</u> = <u>85</u>	<u>65</u> − 2 = <u>63</u>	<u>42</u> − 1 = <u>41</u>
88 − <u>4</u> = <u>84</u>	<u>55</u> − 2 = <u>53</u>	<u>40</u> − <u>1</u> = <u>39</u>
88 − <u>5</u> = <u>83</u>	<u>45</u> − 2 = <u>43</u>	<u>38</u> − <u>1</u> = <u>37</u>
88 − <u>6</u> = <u>82</u>	<u>35</u> − 2 = <u>33</u>	<u>36</u> − <u>1</u> = <u>35</u>
88 − <u>7</u> = <u>81</u>	<u>25</u> − 2 = <u>23</u>	<u>34</u> − <u>1</u> = <u>33</u>

Add and Subtract Two-Digit Numbers, p. 15

1. a. 28 b. 6 c. 11 d. 15 e. 14 f. 22

2. a. 55 b. 51 c. 67 d. 48 e. 29 f. 66

3. a.	b.	c.
35 + 20 = 55	40 + 17 = 57	33 − 20 = 13
76 + 30 = 106	30 + 33 = 63	78 − 50 = 28
22 + 50 = 72	56 − 20 = 36	99 − 40 = 59

4.
a.
tens	ones
4	2
2	4
6	6
(+)

b.
tens	ones
5	3
	6
5	9
(+)

c.
tens	ones
2	5
5	3
7	8
(+)

d.
tens	ones
3	5
	4
3	9
(+)

5.
a.
tens	ones
9	5
2	0
7	5
(−)

b.
tens	ones
5	8
2	6
3	2
(−)

c.
tens	ones
2	5
	3
2	2
(−)

d.
tens	ones
7	9
6	4
1	5
(−)

6. a. 17 + 21

1	7
2	1
3	8
(+)

b. 34 + 14

3	4
1	4
4	8
(+)

c. 51 + 7

5	1
	7
5	8
(+)

d. 32 + 5

3	2
	5
3	7
(+)

7. a. 24 b. 13 c. 22 d. 12

8.

a.	tens	ones		b.	tens	ones		c.	tens	ones
	2	4			2	3			2	6
+	1	3		+	2	5		+	4	1
	3	7			4	8			6	7

d.	tens	ones		e.	tens	ones		f.	tens	ones
	2	4			1	5			3	4
+		4		+	2	1		+	4	0
	2	8			3	6			7	4

9. a. 50, 80, 80   b. 100, 60, 30   c. 20, 20, 50

10.

a. 57 – 21

	5	7
–	2	1
	3	6

b. 74 – 14

	7	4
–	1	4
	6	0

c. 59 – 7

	5	9
–		7
	5	2

d. 99 – 58

	9	9
–	5	8
	4	1

11.

a. Seventeen fish were put into the freezer.

	2	8
–	1	1
	1	7

b. You have to pay $56.

	2	2
+	3	4
	5	6

c. Mom is 27 years older than John.

	3	8
–	1	1
	2	7

d. Matt has 28 colored pencils now.

	2	2
		6
	2	8

Subtract from Whole Tens, p. 19

1. a. 36, 34, 33, 32   b. 25, 26, 21, 24   c. 48, 42, 47, 44   d. 53, 51, 59, 56

2. a. 70 – 10 – 10 – 10 = 40   b. 90 – 20 – 20 – 20 = 30

3. 10 → 60 → 100 → 80 → 20 → 40 → 10 → 30 → 90 → 60

4. a. 64, 65, 68   b. 42, 43, 44   c. 39, 38, 37   d. 95, 93, 91

5.

a. $10 - 2 = 8$	b. $60 - 5 = 55$	c. $25 - 4 = 21$
$10 + 2 = 12$	$60 + 5 = 65$	$25 + 4 = 29$

6. a. $20 - 16 = \underline{4 \text{ pencils}}$

   b. $17 - 7 = \underline{10 \text{ bushes}}$

   c. $20 - 13 = 7.$ <u>Julie has 7 more stones than Carmen.</u>

     $18 - 13 = 5.$ <u>Jane has 5 more stones than Carmen.</u>

     $13 + 7 = 20.$ <u>Carmen needs seven more stones.</u>

Doubling, p. 21

1. a. $4 + 4 = 8$ b. $6 + 6 = 12$ c. $8 + 8 = 16$

   d. $10 + 10 = 20$ e. $30 + 30 = 60$ f. $50 + 50 = 100$

2.

a.
2	2
+ 2	2
4	4

b.
3	4
+ 3	4
6	8

c.
1	3
+ 1	3
2	6

d.
4	1
+ 4	1
8	2

3.

Double 1 = 2	$6 + 6 = 12$	$11 + 11 = 22$
Double 2 = 4	$7 + 7 = 14$	$12 + 12 = 24$
Double 3 = 6	$8 + 8 = 16$	$13 + 13 = 26$
Double 4 = 8	$9 + 9 = 18$	$14 + 14 = 28$
Double 5 = 10	$10 + 10 = 20$	$15 + 15 = 30$

4. Each child will make 8 sandwiches.

5. You get 6 grapes.

6. Mary moved 8 spaces. Andrea moved 12 spaces.

7. **Each girl gets 4.** First add $5 + 3 = 8$. Eight is double 4.

8.   13 **20** 19 **8** 15 **16**

9.

a. $8 = 4 + 4$	b. $10 = 5 + 5$	c. $4 = 2 + 2$
d. $12 = 6 + 6$	e. $14 = 7 + 7$	f. $16 = 8 + 8$

10.

3	4	5	6	7	8	9	10	11	12	13	14	15
6	8	10	12	14	16	18	20	22	24	26	28	30

11. Each woman will make 10 dolls.

12. Each teacher will get 14 worksheets.

13. <u>You get 5 slices.</u> First add $7 + 3 = 10$.

14. A double batch of brownies makes 32 brownies.

One-Half, p. 24

1. a. 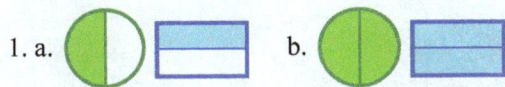 b.

2. a.  b. c. d. e.

3.

a. $5 + 5 = 10$	b. $20 + 20 = 40$	c. $12 + 12 = 24$
$\frac{1}{2}$ of 10 is 5.	$\frac{1}{2}$ of 40 is 20.	$\frac{1}{2}$ of 24 is 12.

4.

$6 + 6 = 12$	$11 + 11 = 22$
$7 + 7 = 14$	$12 + 12 = 24$
$8 + 8 = 16$	$13 + 13 = 26$
$9 + 9 = 18$	$14 + 14 = 28$
$10 + 10 = 20$	$15 + 15 = 30$

$\frac{1}{2}$ of 16 is 8.

$\frac{1}{2}$ of 28 is 14.

$\frac{1}{2}$ of 26 is 13.

$\frac{1}{2}$ of 30 is 15.

$\frac{1}{2}$ of 22 is 11.

5.

a.	b.	c.	d.
$\frac{1}{2}$ of 30 is 15.	$\frac{1}{2}$ of 80 is 40.	$\frac{1}{2}$ of 48 is 24.	$\frac{1}{2}$ of 48 is 24.

6. a. Each one got $30.

   b. 50 students were not sick.

   c. Missie has $10 now.

   d. Mom had 20 apples originally.

$10 + 10 = 20$
$15 + 15 = 30$
$20 + 20 = 40$
$25 + 25 = 50$
$30 + 30 = 60$
$35 + 35 = 70$
$40 + 40 = 80$

Pictographs, p. 26

1. a. <u>Jim</u> rode the most miles. Jim rode <u>70 miles</u>.
   b. The boys that rode the least miles were Greg and <u>Ernest</u>. Greg rode 15 miles. Ernest rode 25 miles.
   c. Matthew rode <u>10 more miles</u> than Dan.
   d. Dan rode <u>25 more miles</u> than Greg.

2.

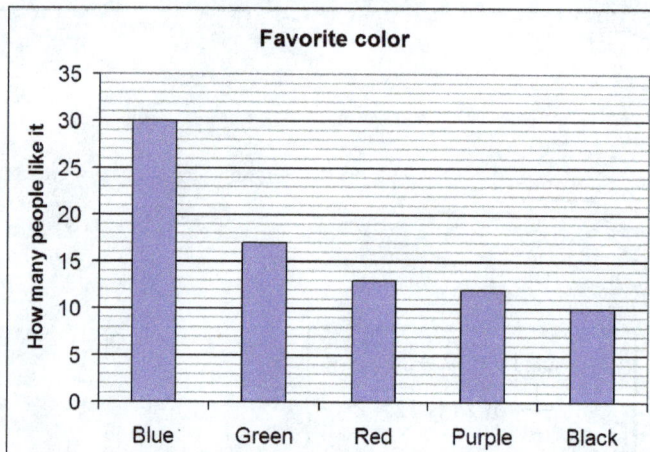

Favorite color

3. a.

	How many?	
oranges	15	
mangos	21	
bananas	24	

    b. 15 + 24 = 39 oranges and bananas together.

    c. Three more bananas than mangos.

4. Answers will vary. Possible questions:

    How many points did Mark get? How many points did Aaron get?
    How many more points did Aaron get than Jack?
    Who got most points? Who got least points?
    How many points did the winner get?
    How many points more did the winner get than Jack?
    How many points did Mark and Aaron get in total?

5. Answers will vary.

Adding with Whole Tens, p. 28

1. a. 64   b. 59 + 20 = 79   c. 34 + 50 = 84
    d. 20 + 13 = 33   e. 40 + 26 = 66   f. 30 + 34 = 64

2. a. 44   b. 10 + 20 + 8 = 38   c. 20 + 20 + 4 = 44   d. 30 + 20 + 1 = 51   e. 50 + 10 + 7 = 67
    f. 40 + 30 + 3 = 73   g. 60 + 20 + 3 = 83   h. 30 + 30 + 7 = 67   i. 70 + 20 + 5 = 95

3. a. 65   b. 47, (20 + 7 + 20)   c. 85, (40 + 5 + 40)   d. 76   e. 66   f. 98   g. 97   h. 98   i. 79

4. Answers will vary. For example, add first 20 and 60 to get 80. Then add 1 to that, to get 81.

5.

5 + 5 = 10	30 + 30 = 60
10 + 10 = 20	35 + 35 = 70
15 + 15 = 30	40 + 40 = 80
20 + 20 = 40	45 + 45 = 90
25 + 25 = 50	50 + 50 = 100

6. She has 15 books left to read. The student can use the chart above to find half of 30.

7. The total cost is $40 + $10 + $20 = $70. Gwen paid $35. The student can use the chart above to find half of 70.

8. He has $61 − $30 = $31 left.

9. a. 10, 40, 30   b. 20, 10, 30   c. 40, 30, 70

10. What do you notice? When you add 10, 20, 30, or 40 to an even number, the answer is also an even number. When you add 10, 20, 30, or 40 to an odd number, the answer is also an odd number.

+ 10	+ 20	+ 30	+ 40
12 22	19 39	32 62	37 77
E E	O O	E E	O O

+ 40	+ 30	+ 20	+ 10
23 63	58 88	7 27	85 95
O O	E E	O O	O O

Puzzle Corner:

Here are two solutions. There are more.

$20 + 30 + 20 = 70$
$+\qquad +\qquad +$
$40 + 40 + 20 = 100$
$+\qquad +\qquad +$
$20 + 30 + 20 = 70$
$=\qquad =\qquad =$
$80\quad 100\quad 60$

$10 + 50 + 10 = 70$
$+\qquad +\qquad +$
$20 + 40 + 40 = 100$
$+\qquad +\qquad +$
$50 + 10 + 10 = 70$
$=\qquad =\qquad =$
$80\quad 100\quad 60$

Subtracting Whole Tens, p. 31

1. a. 20 b. 35 c. 26

2. a. 76, 66, 56, 46, 36, 26, 16   b. 72, 62, 52, 42, 32, 22, 12

3. a. 13, 3   b. 28, 18   c. 46, 26   d. 65, 55   e. 21, 11   f. 41, 31

4.

a. $88 - 10 = 78$	b. $100 - 60 = 40$	c. $34 - 10 = 24$
$88 - 20 = 68$	$90 - 50 = 40$	$44 - 20 = 24$
$88 - 30 = 58$	$80 - 40 = 40$	$54 - 30 = 24$
$88 - 40 = 48$	$70 - 30 = 40$	$64 - 40 = 24$
$88 - 50 = 38$	$60 - 20 = 40$	$74 - 50 = 24$
$88 - 60 = 28$	$50 - 10 = 40$	$84 - 60 = 24$
$88 - 70 = 18$	$40 - 0 = 40$	$94 - 70 = 24$

5. a. 30 kg + 18 kg + 20 kg = 68 kg.
   b. $30 + $30 + $30 = $90. No, he cannot.
   c. The books cost $30 in total. $50 – $30 = $20. He has $20 left.

Puzzle corner: There are many solutions for both.

$40 + 50 = 90$
$-\qquad -$
$10 + 20 = 30$
$=\qquad =$
$30\quad 30$

$50 - 10 = 40$
$+\qquad +$
$30 - 0 = 30$
$=\qquad =$
$80\quad 10$

$30 + 60 = 90$
$-\qquad -$
$0 + 30 = 30$
$=\qquad =$
$30\quad 30$

$45 - 5 = 40$
$+\qquad +$
$35 - 5 = 30$
$=\qquad =$
$80\quad 10$

Review: What numbers make 10?	$1 + \underline{9} = 10$ $7 + \underline{3} = 10$ $4 + \underline{6} = 10$	$8 + \underline{2} = 10$ $5 + \underline{5} = 10$ $9 + \underline{1} = 10$	$3 + \underline{7} = 10$ $6 + \underline{4} = 10$ $2 + \underline{8} = 10$

1. a. $33 + 7 = 40$ b. $43 + 7 = 50$ c. $27 + 3 = 30$ d. $36 + 4 = 40$ e. $62 + 8 = 70$ f. $54 + 6 = 60$

2.

a. $\underline{10}$, 13, $\underline{20}$	b. $\underline{50}$, 57, $\underline{60}$	c. $\underline{40}$, 46, $\underline{50}$
d. $\underline{80}$, 81, $\underline{90}$	e. $\underline{70}$, 78, $\underline{80}$	f. $\underline{90}$, 94, $\underline{100}$

3. a. $56 + 4 = 60$ b. $35 + 5 = 40$ c. $49 + 1 = 50$

4.

a. $3 + \underline{7} = 10$ $23 + \underline{7} = 30$	b. $4 + \underline{6} = 10$ $44 + \underline{6} = \underline{50}$	c. $7 + \underline{3} = 10$ $17 + \underline{3} = \underline{20}$

5. a. 7   b. 9   c. 6   d. 2   e. 6   f. 4   g. 3   h. 5   i. 8   j. 2   k. 9   l. 1

6.

a. $36 + 4 = 40$ $40 - 4 = 36$	b. $57 + 3 = 60$ $60 - 3 = 57$	c. $83 + 7 = 90$ $90 - 7 = 83$
d. $66 + 4 = 70$ $70 - 4 = 66$	e. $95 + 5 = 100$ $100 - 5 = 95$	

7. a. $30 + 7 + 3 = 40$. <u>Jeanine needs three more dollars.</u>
   b. $20 + 10 + 10 = 40$. <u>Derek needs ten more dollars.</u>
   c. $12 + 20 + 8 = 40$. <u>Muhammad needs eight more dollars.</u>

Puzzle corner. Answers will vary because there are many possible solutions. These are just two example solutions.

100	−	10	−	50	= 40
−		+		+	
30	+	30	+	30	= 90
=		=		=	
70		40		80	

100	−	30	−	30	= 40
−		+		+	
30	+	10	+	50	= 90
=		=		=	
70		40		80	

Going Over Ten, p. 36

1. a. 14   b. 17   c. 18   d. 13

2.

a. 7 + 8 = 15	b. 8 + 8 = 16	c. 6 + 5 = 11
d. 9 + 4 = 13	e. 8 + 5 = 13	f. 8 + 9 = 17
g. 7 + 7 = 14	h. 9 + 9 = 18	

3. a. 13 + 9 = 22   b. 15 + 8 = 23   c. 17 + 7 = 24
   d. 24 + 7 = 31   e. 25 + 6 = 31   f. 37 + 9 = 46
   g. 36 + 6 = 42   h. 48 + 4 = 52   i. 58 + 5 = 63

4.

a.   28 + 8
       / \
 28 + 2 + 6
 30 + 6 = 36
	b. 47 + 5
	       / \
	 47 + 3 + 2
	 50 + 2 = 52
		c. 79 + 9
		       / \
		 79 + 1 + 8
		 80 + 8 = 88

d.   39 + 3
       / \
 39 + 1 + 2
 40 + 2 = 42
	e. 27 + 5
	       / \
	 27 + 3 + 2
	 30 + 2 = 32
		f. 38 + 7
		       / \
		 38 + 2 + 5
		 40 + 5 = 45

5. a. 40   b. 42   c. 64   d. 35   e. 62   f. 61

6. a. 39   b. 40   c. 52   d. 38   e. 59   f. 62

7. a.

		Count
Dad	||||| ||||| ||||| ||||	19
Mom	||||| ||||| ||||| ||||| ||||| |||	28
Mary	||||| ||||| ||	12
Mark	||||| ||||| ||||| ||||| |||||	25
Angie	||||| ||||| ||||| ||||| ||||| ||||| |||||		36

b.

How many birds family members saw

Dad	Mom	Mary	Mark	Angie
19	28	12	25	36

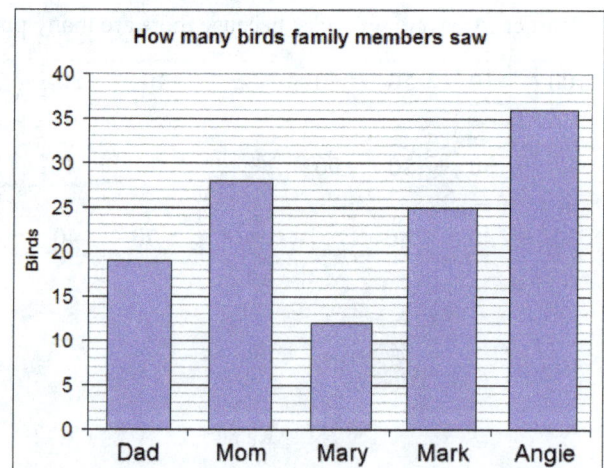

c. Dad saw seven more
   birds than Mary.
d. Angie saw 11 more
   birds than Mark.

```
   3 6
 - 2 5
 -----
   1 1
```

Add with Two-Digit Numbers Ending in 9, p. 40

1.

1. a. 19 + 5 = <u>24</u>	b. 29 + 7 = <u>36</u>	c. 49 + 5 = <u>54</u>
d. 29 + 8 = <u>37</u>	e. 39 + 6 = <u>45</u>	f. 49 + 9 = 58

2. a. 19 + 7 = <u>26</u>   b. 49 + 3 = <u>52</u>   c. 39 + 4 = <u>43</u>
      <u>9</u> + <u>7</u> = <u>16</u>   <u>9</u> + <u>3</u> = <u>12</u>   <u>9</u> + <u>4</u> = <u>13</u>

3. a. 12, 22 b. 15, 45 c. 13, 53 d. 16, 46, 36 e. 18, 78, 88   f. 14, 24, 64
   The answers to the problems in each box end in the same digit.

4. a. 5, 6, 4, 9, 8, 7   b. 13, 17, 14, 15, 18, 16
   c. 7, 9, 4, 6, 5, 8   d. 15, 13, 14, 11, 17, 12

5. a. 12   b. 14   c. 11

6.

a. +1 +2 +3 +4 +5 +6 +7 +8
   0   1   3   6   10   <u>15</u>   <u>21</u>   <u>28</u>   <u>36</u>

b. +4 +4 +4 +4 +4 +4 +4 +4
   <u>24</u>   <u>28</u>   <u>32</u>   <u>36</u>   <u>40</u>   44   48   52   56

Add a Two-Digit Number and a Single-Digit Number Mentally, p. 42

1.

a. 18 + 6 = <u>24</u>	b. 28 + 7 = <u>35</u>	c. 48 + 8 = <u>56</u>
d. 38 + 4 = <u>42</u>	e. 38 + 6 = <u>44</u>	f. 48 + 5 = <u>53</u>

2.

a. 18 + 7 = <u>25</u>	b. 38 + 6 = <u>44</u>	c. 58 + 5 = <u>63</u>

3. a. 11, 21 b. 14, 44 c. 12, 82   d. 10, 40, 30 e. 17, 77, 87 f. 13, 23, 63
   The answers to the problems in each box end in the same digit.

89

4. a. 13, 33   b. 14, 84   c. 14, 94   d. 13, 43   e. 15, 35   f. 15, 55

5. To add 73 + 8, I can use the helping problem 3 + 8 = 11 . Then since
   the answer to that is 1 more than 10, the answer to 73 + 8 is 1 more than 70 .

6. 42   b. 54   c. 64

7. a. <u>She needs 14 more eggs.</u> 10 + 14 = 24, or 24 – 10 = 14
   b. <u>They have eaten 20 kilograms of potatoes.</u> 25 – 5 = 20 OR 5 + 20 = 25

Regrouping with Tens, p. 44

1. a.   33   b.   25   c.   38   d.   27   e.   36   f.   25
       + 9   + 8   + 9   + 7   + 18   + 27
      ─────   ─────   ─────   ─────   ─────   ─────
        42   33   47   34   54   52

2. a.   13   b.   24   c.   35   d.   24   e.   44   f.   26   g.   25   h.   39
       + 29   + 38   + 19   + 47   + 17   + 36   + 55   + 35
      ─────   ─────   ─────   ─────   ─────   ─────   ─────   ─────
        42   62   54   71   61   62   80   74

3. a. 57 b. 72 c. 91 d. 98 e. 53 f. 93 g. 57 h. 90 i. 87 j. 95

4. a. 79 b. 87 c. 66 d. 87 e. 74

5. a. 13 + 9 + 11 = 33

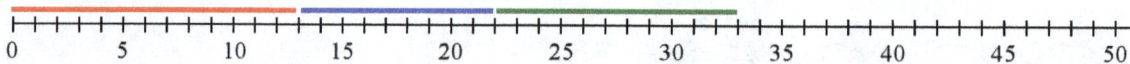

   b. 27 + 16 = 43

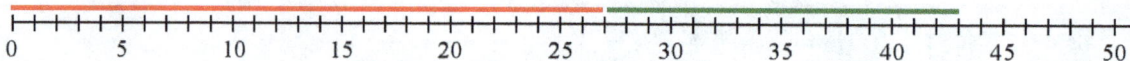

Add in Columns Practice, p. 47

1. a. 80   b. 91   c. 91   d. 70   e. 52   f. 78   g. 116   h. 94   i. 85   j. 49
2. a. 72   b. 64   c. 72   d. 70   e. 94   f. 62   g. 56   h. 74   i. 26   j. 66
3. a. 107   b. 142   c. 112   d. 146
4. a. 128   b. 155   c. 120   d. 147   e. 123   f. 109   g. 126   h. 147   i. 149   j. 137
5. a. <u>43 hours.</u> Add in columns 27 + 16 = 43.
   b. <u>13 more comic books.</u> Subtract 29 - 16 = 13.
   c. <u>The first vase has 2 more flowers than the second.</u> First add 13 + 11 = 24. Then figure out the
      difference between 24 and 22 flowers - it is 2 flowers.
   d. <u>Caleb now has more money, $3 more.</u> Add $24 + 20 = $44. Then figure out the difference of $44 and $41 -- it is $3.
   e. <u>His bill is $43.</u> Add $13 + $9 + $21 = $43.

Mental Addition of Two-Digit Numbers, p. 50

1.

a. 20 + 34 = 54	b. 70 + 18 = 88	c. 50 + 27 = 77
20 + 30 + 4	70 + 10 + 8	50 + 20 + 7

2. a. 27, 36, 52   b. 36, 64, 87   c. 64, 83, 75
3. a. 15, 25, 45   b. 13, 23, 53   c. 12, 52, 82   d. 16, 66, 46

4.

a. $13 + 18 = 3 + 8 + 10 + 10 = 31$	b. $15 + 15 = 5 + 5 + 10 + 10 = 30$
c. $17 + 18 = 7 + 8 + 10 + 10 = 35$	d. $19 + 15 = 9 + 5 + 10 + 10 = 34$
e. $18 + 12 = 30$	f. $13 + 16 = 29$
g. $16 + 17 = 33$	h. $17 + 15 = 32$

5. a. Eight cats live outside.   b. She needs 12 more pounds of cat food.

6. 42, 45, 48, 51, 54, 57, 60, 63, 66

7.

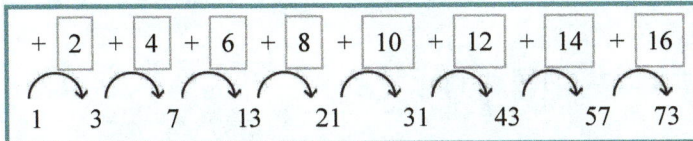

8.

a. $36 + 22$
 $30 + 20$ + $6 + 2$
    50 $+ 8$ $= 58$
	b. $72 + 18$
	 $70 + 10$ + $2 + 8$
	    80 $+ 10$ $= 90$

c. $54 + 37$
 $50 + 30$ + $4 + 7$
    $80 + 11$ $= 91$
	d. $24 + 55$
	 $20 + 50$ + $4 + 5$
	    $70 + 9$ $= 79$

e. $36 + 36$ $30 + 30$ + $6 + 6$ $60 + 12$ $= 72$	f. $42 + 68 = 110$
g. $45 + 18 = 63$	h. $37 + 58 = 95$

Puzzle corner:

a.
```
  2 7
+ 1 4
-----
  4 1
```
b.
```
  6 8
+   3
-----
  7 1
```
c.
```
  2 6
+ 2 5
-----
  5 1
```
d.
```
  1 3
+ 7 8
-----
  9 1
```
e.
```
  3 5
+ 2 6
-----
  6 1
```

Adding Three or Four Numbers Mentally, p. 53

1. a. 24   b. 22   c. 22   d. 21   e. 18   f. 16

Teaching box:

$7 + 5 + 3 + 6 =$ $10 + 11 = 21$	$6 + 9 + 8 + 5 =$ $15 + 13 = 28$	$9 + 8 + 8 + 4 =$ $16 + 4 + 9 = 29$

2.

a. $8 + 8 + 2 + 8$ $= 16 + 10$ $= 26$	b. $7 + 5 + 5 + 6$ $= 13 + 10$ $= 23$	c. $4 + 7 + 2 + 5$ $= 9 + 9$ $= 18$
d. $6 + 7 + 9 + 8$ $= 15 + 15$ $= 30$	e. $8 + 5 + 2 + 6$ $= 10 + 11$ $= 21$	f. $4 + 5 + 3 + 9$ $= 9 + 12$ $= 21$

3. a. 18   b. 24   c. 31   d. 23   e. 20   f. 19   g. 16   h. 21   i. 22

4. Madison took 25 photos of her friends.

5. 7 + 9 = 16, so <u>yes they can share them equally. Each boy will get 8 cars.</u>

6. 8 + 11 = 19 which is an odd number so, **no they cannot share them equally.**

7. a. 80   b. 71   c. 54   d. 47

8. a. <   b. =   c. =   d. <

Adding Three or Four Numbers in Columns, p. 55

1. a. 17   b. 18   c. 23

2. a. 93   b. 92   c. 98   d. 82   e. 97   f. 96   g. 89   h. 80   i. 107   j. 117   k. 119   l. 120

3.

2 1 7 1 7 + 4 9 ——— 8 3 a. Two shirts for $17 each; a pair of jeans for $49.	2 1 7 1 7 + 1 7 ——— 5 1 b. Three buckets for $17 each.	2 2 4 2 4 1 9 + 1 9 ——— 8 6 c. Two shovels for $24 each; two rakes for $19 each.
1 8 9 1 2 + 7 ——— 1 0 8 d. A phone for $89, a phone cover for $12, and chocolate for $7.	1 3 6 3 6 2 3 + 2 3 ——— 1 1 8 e. Two adult tickets for $36 each and two child tickets for $23 each.	3 1 9 2 9 2 9 + 2 9 ——— 1 0 6 f. A child meal for $19 and three adult meals for $29 each.

4. The person does not regroup (carry) but instead writes the sum of the ones directly under the line, and then adds the tens.
   <u>Right answers: a. 81   b. 94</u>

5. a. <u>13 more people.</u> 35 − 22 = 13
   b. <u>Originally there were 6 people.</u> 6 + 13 = 19
   c. No, because 33 + 9 is 42, which is more than 40.
   d. <u>You need 2 buses for 76 people</u> because 40 + 40 = 80, and 80 is more than 76.
      <u>You need 3 buses for 99 people,</u> because 40 + 40 = 80 is not enough, but 40 + 40 + 40 = 120 is enough.
   e. <u>Now there are 34 people.</u> 40 − 6 = 34
   f. <u>There were 30 people left on the bus.</u> 40 − 3 − 7 = 30

6. a. 100   b. 110   c. 113   d. 115

7.

Number	Even?	Odd?	As a double:
8	X		4 + 4
16	X		8 + 8
100	X		50 + 50
19		X	

Number	Even?	Odd?	As a double:
18	X		9 + 9
24	X		12 + 12
15		X	
21		X	

Puzzle Corner:

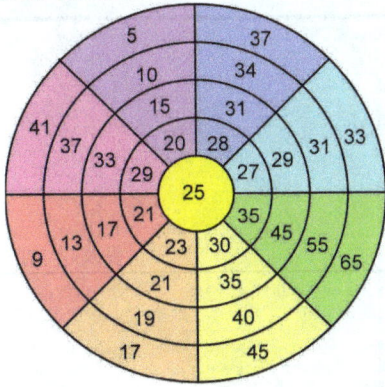

Regrouping in Subtraction, Part 1, p. 59

1.

a. 3 tens 0 ones ⇨ 2 tens 10 ones	b. 3 tens 6 ones ⇨ 2 tens 16 ones
c. 5 tens 1 ones ⇨ 4 tens 11 ones	d. 5 tens 5 ones ⇨ 4 tens 15 ones
e. 4 tens 0 ones ⇨ 3 tens 10 ones	f. 3 tens 7 ones ⇨ 2 tens 17 ones

2. a. What is left? _2_ tens _8_ ones
   b. 5 tens 4 ones → 4 tens 14 ones. What is left? _2_ tens _7_ ones.
   c. 4 tens 3 ones → 3 tens 13 ones. What is left? _1_ ten _8_ ones.
   d. 6 tens 1 one → 5 tens 11 ones. What is left?   _1_ ten _7_ ones.

3.

a. 5 tens 5 ones ⇨	4 tens	15 ones	b. 7 tens 2 ones ⇨	6 tens	12 ones
–	3 tens	7 ones	–	3 tens	5 ones
	1 ten	8 ones		3 tens	7 ones
c. 6 tens 0 ones ⇨	5 tens	10 ones	d. 6 tens 4 ones ⇨	5 tens	14 ones
–	2 tens	7 ones	–	3 tens	8 ones
	3 tens	3 ones		2 tens	6 ones
e. 7 tens 6 ones ⇨	6 tens	16 ones	f. 5 tens 0 ones ⇨	4 tens	10 ones
–	4 tens	7 ones	–	2 tens	2 ones
	2 tens	9 ones		2 tens	8 ones
g. 8 tens 1 one ⇨	7 tens	11 ones	h. 6 tens 3 ones ⇨	5 tens	13 ones
–	6 tens	5 ones	–	2 tens	8 ones
	1 tens	6 ones		3 tens	5 ones

4. a. Jessica now has 13 pencils. 27 – 10 – 4 = 13.
   b. Three more pencils. 13 – 10 = 3.   c. Nine more pencils. 13 – 4 = 9.

93

Regrouping in Subtraction, Part 2, p. 62

1.

a. 6 tens 0 ones → 5 tens 10 ones Take away 3 tens, 9 ones. <table><tr><td>6</td><td>0</td></tr><tr><td>− 3</td><td>9</td></tr><tr><td>2</td><td>1</td></tr></table>	b. 7 tens 1 one → 6 tens 11 ones Take away 1 ten, 6 ones. <table><tr><td>7</td><td>1</td></tr><tr><td>− 1</td><td>6</td></tr><tr><td>5</td><td>5</td></tr></table>
c. 3 tens, 5 ones → 2 tens 15 ones Take away 1 ten, 7 ones. <table><tr><td>3</td><td>5</td></tr><tr><td>− 1</td><td>7</td></tr><tr><td>1</td><td>8</td></tr></table>	d. 8 tens → 7 tens 10 ones Take away 3 tens, 4 ones. <table><tr><td>8</td><td>0</td></tr><tr><td>− 3</td><td>4</td></tr><tr><td>4</td><td>6</td></tr></table>
e. 7 tens, 6 ones → 6 tens 16 ones Take away 4 tens, 8 ones. <table><tr><td>7</td><td>6</td></tr><tr><td>− 4</td><td>8</td></tr><tr><td>2</td><td>8</td></tr></table>	f. 9 tens → 8 tens 10 ones Take away 5 tens, 1 one. <table><tr><td>9</td><td>0</td></tr><tr><td>− 5</td><td>1</td></tr><tr><td>3</td><td>9</td></tr></table>
g. 5 tens, 4 ones → 4 tens 14 ones Take away 2 tens, 5 ones. <table><tr><td>5</td><td>4</td></tr><tr><td>− 2</td><td>5</td></tr><tr><td>2</td><td>9</td></tr></table>	h. 8 tens → 7 tens 10 ones Take away 4 tens, 7 ones. <table><tr><td>8</td><td>0</td></tr><tr><td>− 4</td><td>7</td></tr><tr><td>3</td><td>3</td></tr></table>
i. 7 tens, 4 ones → 6 tens 14 ones Take away 3 tens, 8 ones. <table><tr><td>7</td><td>4</td></tr><tr><td>− 3</td><td>8</td></tr><tr><td>3</td><td>6</td></tr></table>	j. 4 tens 7 ones → 3 tens 17 ones Take away 2 tens, 9 ones. <table><tr><td>4</td><td>7</td></tr><tr><td>− 2</td><td>9</td></tr><tr><td>1</td><td>8</td></tr></table>

2. In this exercise the child is asked to check each subtraction with addition. Adding back is just one method of checking. Estimating the result beforehand is another method of checking - it will not tell you if you got it exactly right but it does check that the result is at least reasonable. Yet another way is to solve the same problem with some different method (there might not always be a "different method"). It is not good that children get used to always checking answers for math problems from the back of the book or from their teacher. They should get used to checking the result themselves. You can encourage that even if the problem set does not tell them to, to check the answers.

a. 29 b. 62 c. 27 d. 55. e. 57 f. 26 g. 38 h. 39 i. 22. j. 38 k. 13 l. 38

Puzzle corner:

9	3
− 1	8
7	5

8	3
− 6	7
1	6

8	0
− 3	8
4	2

8	2
− 1	4
6	8

6	2
−	9
5	3

Regrouping in Subtraction, Part 3, p. 65

1. a. no b. yes c. yes

2. a. YES; 44 b. NO; 25 c. YES; 26 d. YES; 52
   e. NO; 72 f. YES; 24 g. NO; 10 h. YES; 19

3. a. 44, 48 b. 6, 5 c. 92, 4

4. a. $45 - 27 = 18$ b. $34 - 13 = 21$ c. $31 - 9 - 8 = 14$

5. a. $22 - 9 = 13$

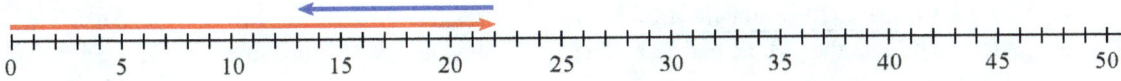

   b. $36 - 12 = 24$

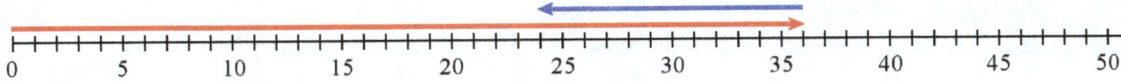

   c. $44 - 17 = 27$

6. a. 94 b. 57, 82 c. 12, 61 d. 42, 99 e. 37, 60 f. 18, 66 g. 13, 54 h. 46, 85

7. a. Emily picked 16 rows more.
   b. Peter sold 27 more tickets.
   c. They sold 97 tickets.
   d. The doll with the pink dress costs $40.
   e. The total cost was $85.

Word Problems, p. 69

1. a. Andy has three more goats than Isabella.
   b. Sandy has 23 goats.
   c. Henry has ten goats.
   d. Daniel has 43 cows.

2. a. Mason has 22 tennis balls.
   b. Logan has 34 tennis balls.
   c. The yellow dress costs $25.
   d. Together, the blue and yellow dresses would cost $66.

3. a. 2 b. 22 c. 8 d. 7 e. 64 f. 95 g. 51 h. 11 i. 42

4. a. $11 - 8 = 3 + 3 = 6$ km   <u>Zoe rode 6 km less than Ryan for 2 days.</u>
   b. $32 - 8 = 24 + 32 = 56$ <u>In total, the two girls have 56 dolls.</u>

Puzzle corner:

   a. △ = 8 b. △ = 26 c. △ = 50   d. -f. Answers will vary.

Mental Subtraction, Part 1, p. 72

1. a. $(51 - \underline{1}) - 4 = 46$ b. $(62 - \underline{2}) - 5 = 55$ c. $(33 - \underline{3}) - 1 = 29$ d. $(92 - \underline{2}) - 3 = 87$
   e. $(75 - \underline{5}) - 1 = 69$ f. $(63 - \underline{3}) - 4 = 56$ g. $(35 - \underline{5}) - 2 = 28$ h. $(74 - \underline{4}) - 1 = 69$

2. a. 44, 46, 48, 45 b. 38, 37, 39, 36

3. a. 5, 15, 35 b. 9, 19, 29 c. 3, 43, 23 d. 6, 56, 36 e. 5, 25, 85 f. 8, 78, 28

4. a. Amy has $25 now.
   b. Peter needs $5 more.
   c. 47 are red.

5. a. 29, 66   b. 56, 28   c. 43, 87   d. 55, 75

Mental Subtraction, Part 2, p. 74

1. a. 8   b. 2   c. 7   d. 4   e. 8   f. 7
   g. 7   h. 8   i. 7   j. 5   k. 9   l. 7

2. a. Larry has 53 cars. 46 + 7 = 53. Mickey has 39 cars. 46 − 7 = 39.
   b. Andy has $10 now. After he bought the gift, he had $33 − $28 = $5. Then he got $5 more so he had $5 + $5 = $10.

3. a. 56   b. 25   c. 21   d. 23   e. 16   f. 32   g. 33   h. 13   i. 11

4. a. There were 8 more books. 42 − 34 = 8.
   b. They both now have 38 books. The second shelf has now 42 − 4 = 38, and the first shelf has 34 + 4 = 38.

5. 35 − 18 = 17

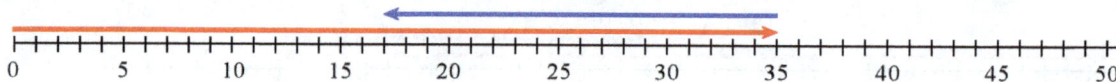

6. a. 14   b. 61   c. 63   d. 26   e. 34   f. 62

7. a. Terry has 43 pages left to read. 95 − 52 = 43.
   b. Now he has 34 pages left to read. You can solve this in many ways. For example: he has now read 52 + 9 = 61 pages. So, he has 95 − 61 = 34 pages left. Or, since he had 43 pages left to read earlier, now he has 43 − 9 = 34 pages left.

8. There are various methods you can use such as breaking a ten into ones to subtract or adding up to the next number or subtracting in parts. If the student cannot answer this question then he or she needs to review what has already been taught in previous lessons.

   a. 16   b. 34

Puzzle corner: a. $\triangle$ = 12   b. $\triangle$ = 9 $\square$ = 13   c. $\triangle$ = 4 $\square$ = 18

Review, p. 77

1. a. 27, 52   b. 36, 87   c. 64, 75

2. a. 35, 61   b. 27, 81   c. 92, 95

3.

a. 36 + 22	b. 72 + 18
30 + 20 + 6 + 2	70 + 10 + 2 + 8
__50__ + __8__ = __58__	__80__ + __10__ = __90__
c. 54 + 37	d. 24 + 55
50 + 30 + 4 + 7	20 + 50 + 4 + 5
__80__ + __11__ = __91__	__70__ + __9__ = __79__

4. a. Ted earned $50. $25 + $25 = $50   Together they earned $75. $50 + $25 = $75
   b. Leah has 12. Half of 24 is 12.

5. a. 71   b. 72   c. 71   d. 61   e. 72
   f. 93   g. 95   h. 121   i. 117   j. 117

6. a. The total cost was $73. $18 + $15 + 40 = $73
   b. The total bill is $102. $34 + $34 + $34 = $102
   c. There are 106 stickers in total. 29 + 29 + 22 + 26 = 106
   d. No, he cannot. $47 + $15 = $62, which is less than $65. He needs $3 more.

More from math MAMMOTH

Math Mammoth has a variety of resources to fit your needs. All are available as economical downloads, and most also as printed copies.

- **Math Mammoth Light Blue Series**
  A complete curriculum for grades 1-7. Each grade level includes two student worktexts (A and B), which contain all the instruction and exercises all in the same book, answer keys, tests, cumulative reviews, and a worksheet maker. International (all metric), Canadian, and South African versions are also available.

  https://www.MathMammoth.com/complete-curriculum

  https://www.MathMammoth.com/international/international

  https://www.MathMammoth.com/canada/

  https://www.MathMammoth.com/south_africa/

- **Math Mammoth Skills Review Workbooks**
  These workbooks are intended to be used alongside the Light Blue series full curriculum, and they provide additional review to the topics studied in the main curriculum, in a spiral manner.
  https://www.MathMammoth.com/skills_review_workbooks/

- **Math Mammoth Blue Series**
  Blue Series books are topical worktexts for grades 1-7, containing both instruction and exercises. The topics cover all elementary mathematics from 1st through 7th grade. These books are not tied to grade levels, and are thus great for filling in gaps.
  https://www.MathMammoth.com/blue-series

- **Make It Real Learning**
  These activity workbooks concentrate on answering the question, "Where is math used in real life?" The series includes various workbooks for grades 3-12.
  https://www.MathMammoth.com/worksheets/mirl/

- **Review Workbooks**
  Workbooks for grades 1-7 that provide a comprehensive review of one grade level of math—for example, for review during school break or summer vacation.
  https://www.MathMammoth.com/review_workbooks/

Free gift!

- Receive over 350 free sample pages and worksheets from my books, plus other freebies:
  https://www.MathMammoth.com/worksheets/free

Lastly...

- Inspire4 is an inspirational website for the whole family I've been privileged to help with:
  https://www.inspire4.com

www.ingramcontent.com/pod-product-compliance
Lightning Source LLC
Chambersburg PA
CBHW051228200326
41519CB00025B/7289